VOLUME SEVEN

ADVANCES IN STEM CELLS AND THEIR NICHES

VOLUME SEVEN

ADVANCES IN STEM CELLS AND THEIR NICHES

Academic Press is an imprint of Elsevier
50 Hampshire Street, 5th Floor, Cambridge, MA 02139, United States
525 B Street, Suite 1650, San Diego, CA 92101, United States
The Boulevard, Langford Lane, Kidlington, Oxford OX5 1GB, United Kingdom
125 London Wall, London EC2Y 5AS, United Kingdom

First edition 2023

ISBN: 978-0-443-19352-1
ISSN: 2468-5097

For information on all Academic Press publications
visit our website at https://www.elsevier.com/books-and-journals

Publisher: Zoe Kruze
Editorial Project Manager: Naiza Ermin Mendoza
Production Project Manager: James Selvam
Cover Designer: Miles Hitchen

Typeset by STRAIVE, India

Contents

Contributors

Livia E. Lisi-Vega
National Health Service Blood and Transplant; Department of Haematology, University of Cambridge; Wellcome Trust – Medical Research Council Cambridge Stem Cell Institute, Cambridge, United Kingdom

Simón Méndez-Ferrer
National Health Service Blood and Transplant; Department of Haematology, University of Cambridge; Wellcome Trust – Medical Research Council Cambridge Stem Cell Institute, Cambridge, United Kingdom

Etienne C.E. Wang
National Skin Centre, Skin Research Institute of Singapore, Singapore, Singapore

CHAPTER ONE

Fueling fate: Metabolic crosstalk in the bone marrow microenvironment

Livia E. Lisi-Vega[a,b,c] and Simón Méndez-Ferrer[a,b,c,*]
[a]National Health Service Blood and Transplant, Cambridge, United Kingdom
[b]Department of Haematology, University of Cambridge, Cambridge, United Kingdom
[c]Wellcome Trust – Medical Research Council Cambridge Stem Cell Institute, Cambridge, United Kingdom
[*]Corresponding author: e-mail address: sm2116@cam.ac.uk

Contents

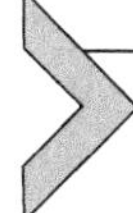

1. Introduction

1.1 The BM stem cell niche: An important piece in the puzzle of hematopoiesis

Hematopoietic stem cells (HSCs) sit at the apex of the organizational hierarchy of the blood (Laurenti & Göttgens, 2018) and it is only through tight regulation of their function and that of their progenitors, that the hematopoietic system can tackle the herculean task of producing the billions of blood cells required daily in an adult (Kaushansky, 2006). Although scarce

Advances in Stem Cells and their Niches, Volume 7
ISSN 2468-5097
https://doi.org/10.1016/bs.asn.2023.08.001

and usually quiescent, HSCs and their output are highly adaptable. This allows HSCs and their progeny to successfully keep up with the daily requirements of the hematopoietic system as well as with the unexpected changes in demand. Like, for example, pathogen infection or chemotherapy treatment, where a need to replenish the blood compartment arises (Baldridge, King, Boles, Weksberg, & Goodell, 2010; Wilson et al., 2008). The precise cues and mechanisms governing the delicate equilibrium between HSCs and their differentiated progeny constitute the focal point of the intricate puzzle that is hematopoiesis in health and disease.

In adults, the bone marrow (BM) is the primary site responsible for blood production (Morrison & Scadden, 2014). In a healthy BM, HSCs interact with and are regulated by a specialized multicellular microenvironment referred to as the HSC niche (Sánchez-Aguilera & Méndez-Ferrer, 2017). This close relationship between blood and bone is a result of millions of years of co-evolution (Kapp et al., 2018; Tirado, Balasundaram, Laaouimir, Erdem, & van Gastel, 2023). In fact, local interactions between HSCs and this BM niche are capable of supporting HSC survival and self-renewal capacity as well as regulating their function. In recent years, the interactions between HSCs (or their malignant counterparts) with the niche have been subject to great scrutiny and are extensively reviewed elsewhere (Kandarakov, Belyavsky, & Semenova, 2022; Tirado et al., 2023; Wei & Frenette, 2018). Instead, this review aims to provide the reader with an overview of the BM niche and the interactions between its cellular components from a metabolic perspective.

While the term "stem cell" can be traced as far back as the 19th century (Dröscher, 2014), the concept of stem cell niche only first appears in 1978, when Schofield proposes a primitive version of the current "niche hypothesis." Building on previous work by Till and McCulloch (Siminovitch, Mcculloch, & Till, 1963; Siminovitch, Till, & McCulloch, 1964), Schofield attributed the disparity observed in the reconstitution capacities of spleen colony-forming assay (CFU-S) derived stem cells and total BM cells to the existence of a stem cell niche capable of nurturing HSC function and regulating hematopoiesis (Schofield, 1978). Fast forward 50 years and this "stem cell niche" has materialized into more precise anatomical locations inside the BM, where macroscopic niches be further subdivided into endosteal and central fractions. The central niche is comprised by the innermost part of the BM and is permeated by arterioles and sinusoids (Méndez-Ferrer et al., 2020). The endosteal niche on the other hand, can be found closer to the bone surface. These two specialized compartments also differ in terms of cell composition. For example, the endosteal niche is characterized by the presence of bone-lining osteoblasts (Lévesque, Helwani, & Winkler, 2010), whereas most sinusoids and HSCs are found in the central BM (Ding, Song, & Luo, 2012). A quick look at the literature soon reveals early reports of researchers looking for a "bona fide" HSC niche and claiming the relevance of either of these endosteal/central niche-associated cell types in orchestrating HSC function (Ambrosi & Chan, 2021; Ellis & Nilsson, 2012).

Advances in high-resolution imaging and the identification of reliable HSC markers have managed to partly reconcile these discrepancies (Kiel et al., 2005; Xie et al., 2009). In fact, HSCs can be found across the central BM and endosteum, where they preferentially locate adjacent to blood vessels. The long-sought after "HSC niche" has therefore been proposed to be perivascular, and to mostly depend on the endothelial cells (ECs) and BMSCs found around sinusoids and arterioles. Perivascular regions are now considered the main anatomical location where HSCs reside, with different subsets of HSCs and progenitors possibly populating specific areas within the BM. The existing experimental evidence seems to suggest that factors such as TGFβ or SCF promotes HSC quiescence in these niches (Nakamura-Ishizu, Ito, & Suda, 2020; Oguro, Ding, & Morrison, 2013). In terms of hematopoietic progenitors, endosteal and central niches have been postulated to harbor early lymphoid restricted progenitors and myeloid-biased HSCs respectively (Pinho et al., 2018). Now a different question arises, does a preference already exist for pre-formed lineage-biased HSCs to locate to the above-mentioned niches, or do these distinct niches instruct a lineage preference in initially unbiased HSCs? Further research is needed to clarify these fundamental questions.

Identifying and dissecting the different anatomical locations and/or niches of the BM has proven difficult. This could be partly attributed to an attempt to directly extrapolate the concept of stem cell niche derived from solid tissues to a fluid tissue like the BM (Sánchez-Aguilera & Méndez-Ferrer, 2017). The BM lacks the clear compartmentalization of other tissues and therefore, associating discrete structures with specific niche-like functions can be rather challenging. Recent work by Kokkaliaris and colleagues delved into this exact question and examined the distribution of HSCs relative to BM components previously described as HSC niches. Strikingly, the authors found no statistical difference between HSC localization and computationally generated dots randomly positioned across the 3D BM space (Kokkaliaris & Scadden, 2020). Their results suggest that HSC location is not actively selected for by the existence of distinct niches, but that is mainly determined by the microanatomical properties of the BM. Conceptually, these microanatomical features could be perhaps closely related to previously-reported structures termed *hemospheres*, which can be found around sinusoidal vessels and are composed by BMSCs, endothelial cells (ECs) and hematopoietic cells, with a particular enrichment of $CD150^+$ CD58- HSCs (Wang et al., 2013). These observations questioning the existence of precise anatomical locations in the BM acting as HSC niches might be a result of experimental approaches based on the idea of a "static" niche, rather than thinking of it as a dynamic entity. In zebrafish, the perivascular niche has been shown to remodel upon HSPC arrival into a pocket-like compartment (Tamplin et al., 2015). Intravital imaging also revealed that HSCs dynamically interact with the niche, exhibiting small oscillations at rest and becoming more migratory upon activation, which enables them to come in contact with larger BM areas (Rashidi et al., 2014). To this day, new discoveries continue to challenge what appear to be dominant concepts in the field. Akin to the co-evolution of blood and bone, our knowledge on hematopoiesis and the niche is slowly advancing to reveal the incredibly complex puzzle that is the BM.

1.2 The bone marrow as a complex organ system

In the BM, an ensemble of cell populations comes together to ensure that homeostasis is always maintained. Rather than doing an in-depth description of each of the cellular components and their reported functions within the BM, this section aims to briefly go over those cell types that have proven metabolically relevant in the niche. A particular focus will be placed on BMSCs, given their importance in the metabolic crosstalk.

1.2.1 Osteoblasts and osteoclasts

Osteoblasts (OBs) and osteoclasts (OCs) are the cells mostly responsible for the maintenance of bone and its remodeling, through the regulation of bone formation and resorption (Bolamperti, Villa, & Rubinacci, 2022). Some of the first experiments looking for a "bona fide" HSC niche pointed towards bone-forming OBs as the main constituents of this niche (Calvi et al., 2003). Long-term HSCs (LT-HSCs) were found in contact with N-cadherin+ OB precursors lining the bone surface, but further work failed to establish a significant effect of N-cadherin deletion in OBs in HSC function and hematopoiesis (Zhang et al., 2003). Despite the lack of experimental confirmation of a direct effect of OBs in the HSC population, it is now clear that OBs synthesize many important factors such as TPO or CSF-1, capable of regulating HSPC quiescence and expansion (Galán-Díez & Kousteni, 2017; Zhang et al., 2023). Along these lines, osteolineage cells have been reported to express EphB4 receptor, which binds to EphrinB2 in HSCs, and might have a direct role in LT-HSC expansion during BM reconstitution (Nguyen, Arthur, & Gronthos, 2016). OBs do not only regulate HSC expansion, but can also limit HSC pool size through osteopontin (OPN) production, given that OPN-KO mice exhibit a 2-fold increase in HSC numbers and acquire an aged phenotype (Guidi et al., 2017; Nilsson et al., 2005).

The bone formation and resorption, which is mediated by OBs and OCs, respectively, have been associated with lymphopoiesis and HSC self-renewal. In fact, ablation of OBs directly leads to reductions in the early lymphoid progenitor pool (Panaroni & Wu, 2013; Shen et al., 2021). The process of bone resorption also brings into play pathways associated with myeloid differentiation and HSPC mobilization (Zhang et al., 2023). In fact, the remodeling of the bone by OBs and OCs activates microdamage and endocrine factors. For example, RANKL expressed by osteocytes initiates osteoclastogenesis. But RANKL also triggers mobilization and expansion of LSK HSPCs (Lin- Sca1+ ckit+) and induction of a myeloid bias (Kollet et al., 2006). Bone also stores >99% of body calcium, which is released by bone resorption. When measuring BM $Ca2^{+}$ concentration, it was discovered that HSCs preferentially reside in regions in which the [$Ca2^{+}$] was significantly higher than the serum calcium concentration (Luchsinger et al., 2019; Yeh et al., 2022). However, our understanding of how bone-derived Ca2+ regulates HSC biology is still limited.

1.2.2 Adipocytes

Adipocytes are the most abundant component of the adult human BM, and from a metabolic standpoint, probably one of the most interesting cell types. The switch from a red marrow, devoid of adipocytes, to a yellow marrow rich in fat tissue by adulthood really highlights how intertwined BM adipocytes are with hematopoiesis regulation. Traditionally, this transition is thought to "physically" exemplify the replacement of a very active state of blood production with one of reduced hematopoietic activity (Kandarakov et al., 2022; Małkiewicz & Dziedzic, 2012). These adipocyte to hematopoietic cell changes brought about by the red-to-yellow marrow transition most probably come with previously unappreciated alterations in BM metabolism. The metabolic effects of varying adipocyte numbers on hematopoiesis are in fact seldomly discussed and may very well be behind some of the discrepancies on the role of adipocytes in promoting or suppressing hematopoiesis.

The existing experimental evidence portrays adipocytes as both negative and positive regulators of hematopoiesis. The observation that the accumulation of adipocytes during aging directly correlates with decreased hematopoiesis, initially endorsed their function as negative regulators of this process. Moreover, upon irradiation, hematopoietic restoration is accelerated in "fatless" mice unable to form adipocytes as well as in mice treated with a PPARγ antagonist (an adipogenesis inhibitor) (Naveiras et al., 2009). Mice with a constitutive deletion of the PPARγ gene also happen to present with severe extramedullary hematopoiesis due to a non-cell autonomous mechanism probably mediated by the CXCL12/CXCR4 axis (Wilson et al., 2018). On the other hand, several studies suggest a role for adipocytes in supporting HSC survival, proliferation and differentiation. Adipocytes produce a series of hematopoietic factors, like SCF, CXCL12, IL-8 or LIF, which are known to directly influence HSC function (Tratwal, Rojas-Sutterlin, Bataclan, Blum, & Naveiras, 2021). Adipocyte-derived SCF has been shown to promote HSC regeneration after irradiation or myeloablation. This effect was found to be adipocyte-specific, since conditional deletion of SCF in endothelial cells (ECs) or OBs did not have the same outcome (Zhou et al., 2017). These effects of adipocyte-derived SCF also hold true in the context of metabolic stress. When BM adipocytes lack Kitl, the hematopoietic compartment fails to adapt to metabolic challenges (Zhang et al., 2019). Interestingly, SCF secretion by brown adipose tissue (BAT) has been shown to respond to systemic changes in temperature and food availability. This clearly suggests that in the BM, adipocytes and their secreted factors are probably highly sensitive to both systemic and local metabolic regulatory mechanisms, which in turn influence hematopoiesis.

1.2.3 Endothelial cells

As an important component of the (*peri*)vascular HSC niche, endothelial cells (ECs) have been linked to various HSC-supporting roles. A potential role of endothelial cells as regulators of HSC function was suggested after conditional deletion of the cytokine receptor gp130, which resulted in reduced HSC numbers and BM hypocellularity, together with extramedullary hematopoiesis (Morrison & Scadden, 2014; Yao, Yokota, Xia, Kincade, & McEver, 2005). However, the ability of ECs to regulate hematopoiesis in the BM was not fully recognized until the appearance of in vitro studies looking at the ability of EC/HSC coculture to promote LT-HSC expansion and repopulating capacity (Butler et al., 2010; Li, Johnson, Shelley, & Yoder, 2004). These studies were followed by efforts to dissect the phenotypical characteristics of sinusoidal (sECs) vs. arteriolar ECs (aECs). While both types of ECs produce SCF and Notch ligands, aECs have been more strongly linked to the expansion and proliferation of HSCs through β3 integrin signaling and synthesis of Del-1 (Mitroulis et al., 2017). Overall, aECs-derived SCF and other factors seem to be critical in mediating hematopoietic regeneration. In contrast to aECs, sECs only synthesize a small fraction of the total EC-derived SCF but happen to produce large amounts of other biologically relevant ligands such as E-selectin or CXCL12. Mice deficient in E-selectin have more quiescent HSCs which are resistant to irradiation. When it comes to studying the role of ECs in the perivascular niche, recent findings suggest that ECs cannot be simply classified into two functional categories (arteriolar vs. sinusoidal) that are solely based on their anatomical location. For example, there is a subset of highly proliferative ECs in the BM that locates to the capillaries and expresses Apelin (Apln+). These Apln+ ECs do not necessarily fall under any of the two aforementioned categories (sEC or aEC) but have been shown to be of great importance for the restoration of the vascular network post-irradiation (Chen et al., 2019).

The metabolic pathways governing EC bioenergetics have been well-described in the context of angiogenesis and as potential cancer targets. However, work on the possible effects of EC-HSC metabolic crosstalk is to this date, still quite sparce. Highly proliferative ECs are required for vessel sprouting and angiogenesis. Therefore, the oxygen supply and consequently, the metabolic state of the BM directly depends on the proliferation, and therefore metabolic, activity of ECs. In a similar manner to HSCs, ECs in the adult vasculature are largely quiescent and rely on glycolysis for ATP production to safeguard their redox balance. This allows ECs to reside in hypoxic regions like the BM without compromising their viability (Li, Sun, & Carmeliet, 2019). Arteriolar vessels are less permeable than sinusoids and thus, HSCs in contact with arterioles are in a "ROS low" state, whereas the increased permeability of sinusoids promotes HSPC activation through the exposure to blood plasma and increased ROS levels (Itkin et al., 2016).

1.2.4 BMSCs: The MVPs of the bone marrow

When it comes to the stromal compartment BM mesenchymal stem cells (BMSCs) are probably at the cusp of the pyramid. BMSCs are a rare population of multipotent stem cells capable of differentiating into other cell types such as adipocytes, OBs or chondrocytes (Pittenger et al., 1999). BMSCs have been shown to reside within perivascular niches in both the endosteal and central BM compartments. It is this tight association with perivascular niches that makes them particularly important in maintaining and regulating HSCs (Méndez-Ferrer et al., 2010). BMSCs found in periarterial and perisinusoidal niches differ in their phenotypical characteristics and transcriptional profiles. While those BMSCs located near the arterioles and transition zone vessels highly express Nestin and NG2 (Nes-GFP^{hi} NG2^{hi}), BMSCs around the sinusoids are Nes-GFP^{lo} NG2^{lo} (Kunisaki et al., 2013). Leptin receptor was initially postulated as an exclusive marker of perisinusoidal BMSCs (Ding, Saunders, Enikolopov, & Morrison, 2012), but has since then been shown to be expressed widely inside and out of the BM, including in arteriolar vessels (Méndez-Ferrer et al., 2020; Shen et al., 2021).

In comparison to the differentiation hierarchies and subpopulations of HSCs, little work has been done on defining the subpopulations and differentiation trajectories of BMSCs. In the past, the limitations of bulk RNAseq failed to accurately reflect the cellular heterogeneity of the BMSC population. Until recently, their identification mostly relied on morphology, surface marker expression and in vitro assays which are now known to alter the endogenous cellular properties of BMSCs (Ambrosi et al., 2017; Ambrosi & Chan, 2021; Ghazanfari, Li, Zacharaki, Lim, & Scheding, 2016). Current advances in single cell RNAseq technologies, have now allowed researchers to deconvolute the population landscape of BMSCs and identify functionally-relevant BMSC subgroups (Gao et al., 2021). For instance, in the mouse BM, LepR+ BMSCs consist of a heterogeneous population of multipotent stem cells thought to be the origin of most adipo- and osteo-lineage cells. Tikhonova and colleagues further subdivided these LepR+ BMSCs into four clusters, two of them (P1 and P2) defined as "pro-adipogenic" and two (P3 and P4) representing osteo-primed BMSCs (Tikhonova et al., 2019). The functional importance of some of these subsets like "osteo-primed" BMSCs can be traced back experimentally to previous reports where expression of markers such as Osterix (Osx) in BMSCs and their downstream populations was found to be critical for niche formation and HSC engraftment (Chan et al., 2009; Mizoguchi et al., 2014). Notably, many scRNAseq studies have been hampered by limited gene coverage and have failed to detect lowly expressed genes, such as *Nestin* in BMSCs. However, it is clear that LepR^{+} cells identified later hugely overlap with Nestin-GFP$^+$ cells (Ding, Saunders, et al., 2012; Méndez-Ferrer et al., 2010), overall confirming the key role of BMSCs in the HSC niche. Prior to these single cell RNAseq studies, the importance of these cytokines in HSC regulation was already demonstrated by the fact that deletion of SCF in LepR$^+$ cells results in a reduction of BM HSC numbers. In accordance with these observations, systematic in situ labelling also found that HSPCs tend to locate near sinusoidal vessels adjacent to LepR$^+$ stroma. However, the deletion of CXCL12 in Nes+ BMSCs and not LepR+ BMSCs showed the most prominent effect on the HSC population, leading to the redistribution of HSCs away from arterioles in the BM (Asada et al., 2017). These results further underscore the conclusion that LepR does not mark a different cell population, but overall adds another useful marker to the previously identified HSC-niche forming BMSCs.

Recent studies have further expanded our knowledge of the BM compartment by utilizing a multiOMICs approach. By combining transcriptomics with the BM proteomic secretome, Passaro and colleagues manage to further characterize the stromal components of the BM, classifying them into relevant subpopulations based on cell identity and function (Passaro et al., 2021). This approach shed some light into the metabolic characteristics of BMSCs. Combining their data with that of Tikhonova and colleagues, allowed the authors to extract highly expressed genes in LepR+ Nes^{low} cells and Nes^{high} cells. More precisely, LepR+ BMSCs were found to distribute into clusters that highly expressed genes related to skeletal function, cell-matrix interaction and bone and cartilage formation (Cluster 1 and 3). In contrast with LepR+ BMSCs, Nes^{high} BMSCs belonged to clusters enriched for genes involved in neuronal processes, mitochondrial function and anabolic processes such as RNA processing and translation (Clusters 6,7 and 8). This work highlights the possible functional heterogeneity of BMSCs. For example, it's possible that Nes^{high} enriches for a BMSC subpopulation with a prominent role in the metabolic regulation of HSCs. Nonetheless, overall LepR+ BMSCs are also not oblivious to metabolic changes, and in fact, have been shown to also adapt and reshape the niche in response to systemic metabolism (Fan et al., 2017; Ning et al., 2022; Yue, Zhou, Shimada, Zhao, & Morrison, 2016).

1.2.5 Other cell types in the BM

For the sake of brevity, this introduction to the components of the BM niche has focused on those cell types that give rise to the main architecture of the BM at tissue level, i.e., bone, vessels, fat, etc. However, it is important to bear in mind that the BM niche is also comprised of other cell types, such as megakaryocytes, immune cells or neurons that make up the nerve fibers that innervate the BM. All of which have also been described to interact with HSCs and influence their behavior. Megakaryocytes for example are known to secrete factors that can influence tissue dynamics, e.g., TPO, TGFB1 or CXCL4 (Bruns et al., 2014; Nakamura-Ishizu, Takubo, Fujioka, & Suda, 2014; Zhao et al., 2014). Nerve fibers innervating the BM can be found close to vessels and Nes + BMSCs. Thanks to this privileged location, the nervous system is capable of regulating HSPC localization and BM egress (García-García et al., 2019; Méndez-Ferrer, Lucas, Battista, & Frenette, 2008), and has been shown to even influence HSC quiescence through its effect on stromal cytokine secretion (Fielding et al., 2022).

In terms of immune cells, the BM niche is at the crossroad of blood and immunity (Ambrosi & Chan, 2021). The crosstalk taking place between the different cell types in the BM somehow achieves to create this protective sanctuary for HSCs, while supporting immune cell function (Fujisaki et al., 2011). In fact, a possible T-cell/HSC regulatory axis was recently reported by Shi and colleagues in the setting of multiple sclerosis (MS). Their data suggests a model whereby autoreactive T-cells from the CNS migrate into the BM and induce myeloid-skewing in HSCs via the CCL5/CCR5 axis as a mean to increase the production of myeloid cells that then exacerbate neuroinflammation and MS progression (Shi et al., 2022). Various studies have delineated the significance of certain metabolic pathways in HSC fate or immune cell activation, but there is still a substantial knowledge gap when it comes to how the different metabolic status of HSCs and immune cells might influence one another. The importance of this careful balance between the protection of HSCs and the immuno-inflammatory state of the BM really becomes apparent in disease states, where high inflammation and immune activation have been shown to precipitate the development and progression of various hematological disorders and malignancies. Metabolism largely determines HSC and immune cell states, and therefore, the metabolic crosstalk between HSCs and immune cells is probably an important factor yet to be taken into consideration (Fig. 1).

1.3 The importance of HSC-niche crosstalk in tissue maintenance and disease

The BM niche provides essential signals that control the quiescence, replication, differentiation and mobilization of HSCs and their progeny.

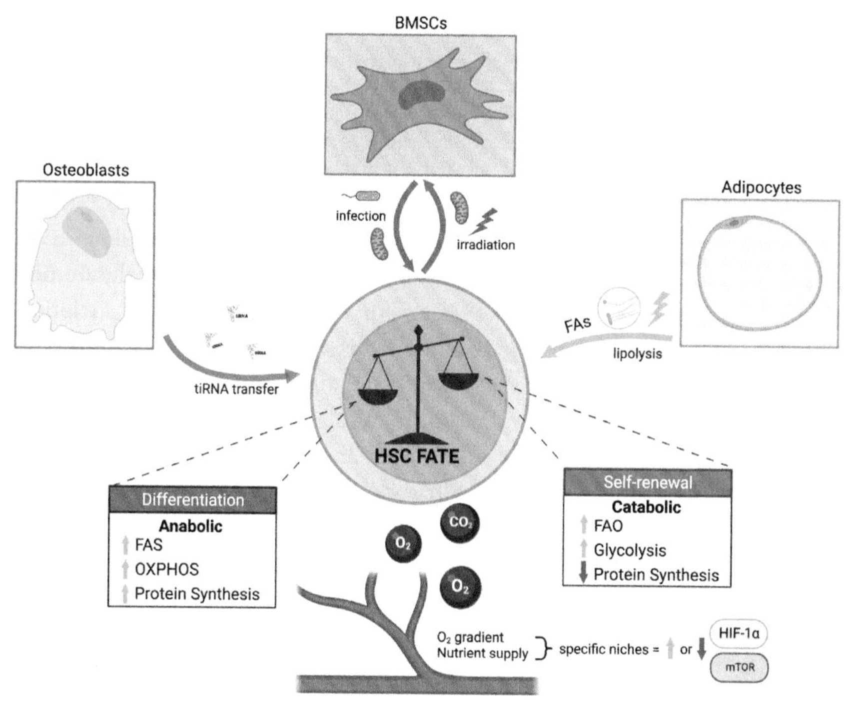

Fig. 1 Overview of the metabolic regulation governing HSC fate and the influence of BM niche metabolic crosstalk on these regulatory networks. The decision to self-renew or proliferate is partly determined by the balance between cellular anabolism and catabolism. HSCs with high self-renewal capacity are typically "metabolically dormant," relying on glycolysis and fatty acid oxidation (FAO), alongside low protein synthesis rates to protect themselves from oxidative stress, damaged organelles, and misfolded proteins to maintain a healthy HSC pool. Conversely, when HSCs undergo differentiation, there is a substantial surge in their energetic demands. This phase is distinguished by a more anabolic profile, driven by heightened levels of protein and fatty acid synthesis, primarily fueled by oxidative phosphorylation (OXPHOS). More importantly, the surrounding niche can significantly influence these metabolic profiles. The niche can modulate critical aspects of HSC metabolism, including nutrient and oxygen supply, lipid availability, mitochondrial levels, and even protein synthesis through the transfer of tiRNAs. In conclusion, the metabolic crosstalk between HSCs and the BM niche plays a pivotal role in shaping HSC fate decisions and regulating homeostasis within the hematopoietic system.

A well-coordinated BM niche is needed to keep HSCs and their output in check and its importance becomes even more apparent in disease. The crosstalk between all BM components is like clockwork, and its perturbation can have big consequences. A large body of work exists on the mechanisms and/or genetic alterations of the BM niche that can lead to disease initiation or promote disease progression. However, there is still a relative paucity of data on the role played by the metabolic crosstalk between HSCs and the niche in disease. Until recently, the extent to which cellular metabolism governs cellular behavior has been heavily overlooked, even in the context of hematological malignancies, where metabolic reprogramming has been more broadly studied.

Alterations of the niche itself can lead to the development of disease. In hematological malignancies, a genetically altered niche can accelerate and promote the transformation of HSCs. For example, ubiquitous deletion of genes such as RARγ, pRB, IkBa or Mib1 in all stromal cells results in abnormal hematopoiesis, BM failure and anemia (Kim et al., 2008; Rupec et al., 2005; Walkley et al., 2007; Walkley, Shea, Sims, Purton, & Orkin, 2007). More accurate dissection of genetic alterations in the stromal population led to the discovery that deletion of an RNA processing enzyme, *Dicer1*, in osteprogenitors (Osx-expressing cells) can give rise to an MDS-like phenotype (Raaijmakers et al., 2010). This phenotype was partly recapitulated by deleting with Osx-Cre the *Sbds* gene in the same population. *Sbds* is mutated in some leukemias as well as Schwachman-Diamond syndrome (SBDS) and interestingly, happens to be regulated by Dicer1. In another striking example of BM niche-driven oncogenesis, the direct expression of a constitutively active form of β-catenin in OBs ($Ctnnb1^{CAosb}$) led to impaired differentiation of hematopoietic progenitors, giving rise to a phenotype that resembled acute myeloid leukemia (AML) (Kode et al., 2014). While these studies confirm the ability of genetic alterations in the BM niche to trigger malignant transformation or that first "oncogenic hit," there seems to be a need for the accumulation of additional mutations in order to induce a full-blown hematopoietic malignancy.

1.3.1 Aging

The changes observed in hematopoiesis with age can be partly attributed to the BM niche. During aging, the central BM niches expand, whereas endosteal niches reduce in size. Since the central and endosteal niches are associated with myelopoiesis and lymphopoiesis respectively, these changes in the BM are thought to favor the lymphoid-to-myeloid skewing observed in elderly patients (Ho et al., 2019; Ho & Méndez-Ferrer, 2020). With age, lymphoid-biased/balanced HSCs seem to acquire a myeloid-like signature but transplanting them into a young recipient can restore their capacity to commit to lymphoid lineages (Montecino-Rodriguez et al., 2019). In light of these findings, it has been suggested that the changes in BM niches happening during aging might favor the expansion of the mutant clones that then drive the appearance of age-related myeloid-malignancies (Méndez-Ferrer et al., 2020).

The aging of the skeletal compartment in the BM also affects hematopoiesis. In general, aged bone and its components have been implicated in the inflammatory state of the BM (Zhang et al., 2023). For example, aged osteoprogenitors exhibit an upregulation of molecules like CSF-1, IL-1 and CCL5 which are known to affect HSC dynamics and promote myeloid differentiation through early activation of transcription factors like PU.1 (Ambrosi et al., 2021; Pietras et al., 2016). Aged BMSCs are primed towards adipocyte differentiation and in mice, BM aging can be mimicked by a high-fat diet (Ambrosi et al., 2017; Solimando, Melaccio, Vacca, & Ria, 2022). With age, adipocytes tend to accumulate in aged BM, negatively impacting hematopoiesis. A higher number of adipocytes implies an overproduction of factors like adipokines, which are known to promote inflammation and as a result, the appearance of an aged-related proinflammatory immune phenotype (Lewis, Edwards, Naylor, & McGettrick, 2021; Yue et al., 2016). Interestingly, Liu et al. recently showed how BM adipocytes with a senescence-associated secretory phenotype (SASP) might be able to trigger secondary senescence in surrounding bone cells and lead to bone deterioration (Liu et al., 2023).

1.3.2 Metabolic disease from a BM perspective

Little attention has been paid so far to how systemic metabolism and metabolic disease might impact the metabolic state of the different cell types in the BM, and how this might alter hematopoietic output. Chronic metabolic conditions such as diabetes usually come hand in hand with a state of systemic inflammation which most likely feeds back to the BM niche. This probably establishes a myeloid-bias in HSCs and in turn perpetuates the chronic inflammation present in these disorders. Diabetic patients are characterized by insulin resistance and impaired insulin secretion, which then leads to a hyperglycemic metabolic state that is accompanied by dyslipidemia and persistent inflammation. The systemic effects of diabetes reach the BM and result in the disruption of the BM niche. With the severity and duration of diabetes more less determining the extent of BM niche remodeling. In general, the BM niche of a diabetic patient exhibits microangiopathy, neuropathy and mobilopathy (impaired stem cell mobilization) (Fadini, Ferraro, Quaini, Asahara, & Madeddu, 2014; Xu, Murphy, & Fleetwood, 2022). The profound changes in the inflammatory milieu and

BM cells associated to diabetes are the culprits behind these structural manifestations of the disease. Although not directly related to BM metabolism, ECs in the BM of diabetic mice were found to produce lower levels of CXCL12. The reduction in EC-derived CXCL12 levels led to the expansion of HSPCs and myelopoiesis (Hoyer et al., 2020). Considering the delicate metabolic balance required to maintain hematopoietic homeostasis in the BM, it remains to be determined whether the alteration in metabolite levels caused by chronic metabolic diseases like diabetes can influence the metabolic crosstalk occurring within the different BM cell types and its downstream effects in hematopoiesis.

The dysregulation of systemic metabolism and the health conditions tied to this dysregulation are increasingly prevalent in Western societies. How these metabolic and inflammatory signals arriving to the BM niche influence the regulatory and metabolic networks within the BM itself is of great significance and still needs to be established.

2. The HSC-niche metabolic axis during normal hematopoiesis

2.1 Hypoxia and glycolysis vs OXPHOS: The duel in HSC fate

Apart from providing energy, metabolic pathways can have tremendous influence on a variety of cellular processes (Etchegaray & Mostoslavsky, 2016). In hematopoiesis we know that metabolism is instrumental in shaping HSC fate decisions and governs the ability of HSCs to respond to different stressors (Nakamura-Ishizu et al., 2020). The factors that determine a cell's metabolism are not just intrinsic to it. A cell's surrounding cell types and signals are of paramount importance when it comes to determining its metabolic status. However, while the metabolic crosstalk between the BM niche and malignant hematopoietic cells has been well characterized, the crosstalk happening with their healthy counterparts has not been subject to as much attention.

2.1.1 The BM and its hypoxic niches

By locating to the BM, HSCs are protected from UV light by cortical bone but are also exposed to a limited oxygen supply (Tirado et al., 2023; Zhang & Sadek, 2014). The unique anatomic structure of the BM, together with the high rate of total oxygen consumption of hematopoietic cells gives rise to the low oxygen tension of the BM (Eliasson & Jönsson, 2010; Mojca, Ro Zman, Ivanovi, & Bas, 1999). Hypoxia was one of the first metabolic properties of the BM niche to be described. Rather than just having extremely low levels of oxygen all throughout the BM, further characterization of the BM architecture suggests that the presence of sinusoids and arterioles probably creates an oxygen gradient within the tissue. Hence, depending on their physical location, HSCs and their progeny might be exposed to different levels of oxygen, and as a result, may differ in metabolism.

The hypoxia marker pimonidazole, revealed that quiescent HSCs tend to locate to more hypoxic regions within the BM which are more distant from capillaries (Kubota, Takubo, & Suda, 2008). In support of these observations were in vitro experiments in which culturing HSCs in low levels of oxygen (1–3%) enhanced their expansion and differentiation as well as their subsequent engraftment potential in vivo (Koller, Bender, Terry Papoutsakis, & Miller, 1992). This need of HSCs for low oxygen levels can be traced back to the main oxygen sensor within the cell and master regulator of metabolism: Hif-1α (Zhang & Sadek, 2014). Hif-1α is stabilized in hypoxic conditions and regulates the expression of hundreds of genes resulting in metabolic adaptation to the lack of oxygen. HSCs stably express high levels of Hif-1α, even under normoxic conditions (Kocabas et al., 2015), and treatment with tirapazamine (a hypoxic toxic reagent) leads to a notable reduction in total HSC numbers (Simsek et al., 2010). Stabilizing or deleting Hif-1α in mice also gives rise to marked HSC functional defects (Takubo et al., 2010). The presence of a narrow range of oxygen concentrations within the BM probably plays a role in determining the specific levels of Hif-1α activation (not too low and not too high) associated with HSC activity (Huang, Chen, Xie, Yu, & Zheng, 2019). In line with this hypothesis, HSCs were shown to rapidly differentiate into progenitor cells as soon as they are exposed to atmospheric levels of oxygen. The cytosolic translocation of the transcription factor TFE3 seemed to be in part responsible for this loss of pluripotency (Wang, Cooper, Broxmeyer, & Kapur, 2022). HSCs collected under normoxic conditions also exhibited reduced mTOR and PI3K activity (Capitano et al., 2021), raising questions regarding the effect that Hif-1α and hypoxia might have on other cellular metabolic pathways. In this regard, it is important to mention that Hif-1α is also regulated by metabolic cues different from hypoxia (Iommarini, Porcelli, Gasparre, & Kurelac, 2017).

2.1.2 Hypoxia, glycolysis and HSCs

The most prominent change that cells experience when switching between normoxic and hypoxic environments is the oxygen that is available to drive aerobic metabolism and the production of energy through OXPHOS. Instead, in hypoxia, ATP levels are maintained through glycolysis, with Hif-1α inducing the expression of key glycolytic enzymes such as LDHA, PKM2 or PDK2. Some of these enzymes, like PKM2, are responsible for directly controlling the rate at which glycolysis occurs (Huang et al., 2019). Hif-1α stabilization also enhances cellular glycolytic flux by promoting PDK activity and inhibiting the entry of pyruvate into the TCA cycle and OXPHOS (Takubo et al., 2013). Deletion of Hif-1α in mice soon highlights the parallelism between the expression of Hif-1α in HSCs and the importance of glycolysis in maintaining HSC quiescence. HSCs lacking Hif-1α quickly switch from glycolysis to mitochondrial respiration and OXPHOS (Takubo et al., 2010), resulting in a loss of quiescence and reconstitution capacity. Similarly, deletion of glycolytic enzymes like LDHA or PKM2 (Hif-1α's downstream targets) also leads to a notable impairment in HSC repopulation capacity (Wang et al., 2014). These findings hint that not only Hif-1α levels, but also glycolysis levels modulate the quiescence and self-renewal capacity of HSCs.

High glycolysis levels might allow HSCs to avoid the cellular stress-associated with residing in the BM's hypoxic environment (e.g., ROS accumulation). Using glycolysis as their main energy source might be key for HSCs. Based on their proteome, LSK HSCs seem to prefer glycolysis in comparison to more differentiated progenitors which appear to use OXPHOS as their main source of energy (Unwin et al., 2006). By rewiring their metabolism in this way, HSCs can shut down all additional energy consumption and unnecessary processes and remain "metabolically quiescent" while maintaining their activity (Arai et al., 2004). These studies are proof of how HSC metabolism influences HSC division balance and can deeply affect the equilibrium between self-renewal and commitment.

2.1.3 Mitochondria, OXPHOS and HSCs

The acquisition of lineage-specific fates by HSCs is now thought to resemble a continuum of cellular states rather than involve major transitions through specific stages. By dynamically switching their metabolic state, quiescent HSCs can fine-tune their fate choices, exit the G0 stage and initiate differentiation (Huang et al., 2019). In this context, the ability of HSCs to quickly switch metabolic profiles (e.g. from glycolysis to OXPHOS) becomes crucial. While glycolysis allows HSCs to produce energy and stay quiescent, cell division and differentiation brings about a steep increase in energy requirements. HSCs suddenly must start producing building blocks for daughter cells and as a result, need energy to fuel macromolecule synthesis and successfully divide. The need for macromolecules directly coming from TCA cycle metabolites and the high demand for ATP can only be met by engaging in mitochondrial respiration. In fact, inhibiting this switch from glycolysis to mitochondrial respiration in HSCs by deleting the mitochondrial phosphatase PTPMT1, impairs differentiation and leads to a concomitant increase in the HSC population (Yu et al., 2013). This could explain why despite relying mostly on glycolysis, HSCs have also been shown to have high levels of mitochondria. These mitochondria, however, possess limited respiratory and turnover capacities. This suggests that perhaps in HSCs, mitochondria are mostly inactive, and only come into play upon differentiation or under stress, when they act as a source of energy (Gan et al., 2010).

However, HSC metabolism is not as straightforward. OXPHOS "fuels" cell cycle entry and HSC commitment but to an extent, is also essential for HSC stemness. Loss of LKB1 in adult HSCs diminishes mitochondrial biogenesis and causes an HSC pool exhaustion phenotype due to their escape from quiescence (Chen et al., 2015). Along these lines, GIMAP5 expression in HSCs was found to be important in maintaining their quiescence partly by sustaining mitochondrial potential levels (Chen et al., 2015). Impairing mitochondrial respiration in adult HSCs through loss of one of the subunits of mitochondrial complex III (RISP) also results in a loss of quiescence and severe pancytopenia with lethality (Ansó et al., 2017). This does not happen in fetal HSCs, where RISP loss just impaired differentiation. The different effects of RISP loss could be explained by the metabolic differences of HSCs in these two developmental stages, with fetal liver HSCs exhibiting higher levels of OXPHOS in comparison to adult HSCs (Manesia et al., 2015). In a similar way to the molecular signatures that have already been described, it is very plausible that even within the BM itself, different subpopulations of HSCs probably exhibit distinct metabolic characteristics. How these metabolic signatures can be identified and how particular "HSC metabolic subpopulations" contribute to HSC homeostasis is yet to be determined.

Work by Ito and colleagues on Tie2^{+} HSCs is probably the closest when it comes to describing a metabolically unique HSC subpopulation. In their paper, the authors define Tie2 as a bona fide marker of top hierarchical HSCs with high self-renewal capacity. These Tie2+ HSCs were characterized by very active fatty acid oxidation (FAO) in comparison to Tie- HSCs and enhanced mitophagy (via increased recruitment of Parkin to mitochondria) (Ito et al., 2016). In this case, the successful removal and degradation of mitochondria was what allowed Tie2+ HSCs to preferentially undergo symmetric cell divisions. The accumulation of damaged mitochondria is therefore a limiting factor when it comes to undergoing self-renewing division. Interestingly, enhancing mitochondrial clearance through deletion of Atad3a in HSCs (which hyperactivates mitophagy) results in an expanded HSPC pool (Jin et al., 2018). These findings emphasize from a different angle how mitochondria and mitophagy regulate HSC maintenance, probably by ensuring that the stem cell daughters do not receive an excess of damaged mitochondria (Filippi & Ghaffari, 2019). The effect of mitophagy on cell fate choice might lie not only in the successful clearance of damaged organelles, but also on its effect in OXPHOS levels. Regardless of the type of cell division, when HSCs initially decide to enter the cell cycle, their metabolic activity must shift from glycolysis to mitochondrial OXPHOS. However, at the end of a symmetrical division, the stem cell daughters need to find a way of returning to glycolysis to maintain their regenerative capacity (Ito, Bonora, & Ito, 2019). Newly active HSCs might achieve this by clearing healthy but active mitochondria to get rid of elevated OXPHOS levels. As a result, mitophagy might play a role in reverting HSCs to a metabolic profile associated with quiescence.

2.1.4 The ambivalence of the BM niche: From hypoxia to mitochondrial transfer

As discussed in the previous sections, the balance between glycolysis and OXPHOS plays an important role in determining the fate of HSCs. Therefore, influencing the energetic profile of HSCs is a powerful tool used by the niche to tightly regulate HSC homeostasis. The physical association of HSC populations with specific niches within the BM has mostly been related to the secretion of soluble factors, while little work has been done on trying to correlate niche preference to the metabolic profile of HSCs. The oxygen gradient of the BM implies that HSCs at different locations are exposed to varying levels of hypoxia, Hif-1a activation and glycolysis. In the case of malignant cells for example, imaging using a metabolic sensor (SONAR) revealed a preference of leukemia initiating cells (LICs) to locate to the more hypoxic endosteal niche, where they can maintain high levels of glycolysis and undergo symmetric division (Hao et al., 2019;

Hu et al., 2021). In normal hematopoiesis, the effect of hypoxia and Hif-1a loss on HSC activity remains controversial. Experiments looking at Meis1, a trans activator of Hif-1a, in HSCs first pointed at intrinsic regulatory networks in HSCs as the main effectors, rather than the hypoxic niche itself (Kocabas et al., 2012). On the other hand, some recent studies propose that the role of Hif-1a in regulating HSCs might be through its influence in the niche and how it "programs" niche components to support hematopoiesis (Vukovic et al., 2015).

A question arises: how exactly does nutrient and oxygen supply to the BM affect the metabolic crosstalk between HSCs and the niche? Glycolysis is key to maintaining HSC quiescence, and the main metabolic pathway fueling immune effector cell function (Buck, Sowell, Kaech, & Pearce, 2017). In the case of immune cells, metabolic adversity is known to compromise their function (Kirkwood, Kapahi, & Shanley, 2000). A few papers analyzing the effect of dietary restriction in immune cell homeostasis found an overall relocation of peripheral immune cells to the BM. The immune cells that re-homed to the BM would then tend to reprogram their metabolism to use lipids instead of glucose as their main source of fuel (Collins et al., 2019; Jordan et al., 2019; Nagai et al., 2019). The exact metabolic crosstalk mediating the return of hematopoietic cells to the BM during caloric restriction and the impact of this new influx of cells on the BM niche has however, not been explored. What really becomes apparent with these studies is that the BM niche acts as an energetic/metabolic "safe harbor," for HSCs and other hematopoietic cell types (Goldberg & Dixit, 2019). To date, not much work can be found on how nutrient restriction or prolonged starvation affects the metabolic equilibrium between HSCs and the BM niche, and how this in turn influences hematopoietic output. So far, experiments looking at calorically restricted mice have reported enhanced HSC quiescence and promotion of HSC regeneration capacity (Tadokoro & Hirao, 2022; Tang et al., 2016). These effects could be directly related to BM niche metabolic regulation and alterations in the glycolysis vs OXPHOS levels of HSCs.

Another important factor in determining the levels of OXPHOS vs. glycolysis in HSCs, is their mitochondrial mass. In this regard, it is important to consider the transfer of mitochondria from niche cells to HSCs (Singh & Cancelas, 2021). This mechanism appears to be of special importance in allowing HSCs to adapt their output in response to stress and regeneration. In the case of bacterial infection, transfer of mitochondria from BMSCs to HSPCs helps induce a shift in HSC metabolism towards OXPHOS, leading to the subsequent expansion and differentiation of HSPCs into leukocytes (Mistry et al., 2019). Interestingly, there is some degree of reciprocity in this type of HSC-niche metabolic crosstalk, and in a transplantation setting, HSPCs were capable of transferring mitochondria to damaged host BMSCs, improving their metabolic recovery and increasing hematopoietic reconstitution (Golan et al., 2020). All in all, either through mitochondrial transfer or its influence on HSC energy and oxygen supply, the niche can easily tip the balance from quiescence to differentiation and vice versa.

2.2 Other metabolic pathways shaping HSC fate

In addition to the balance between glycolysis and OXPHOS, other metabolic pathways also have a card to play when it comes to the maintenance of HSC function (Huang et al., 2019). Metabolically, stem cell fate decisions can be boiled down to the balance between anabolism and catabolism. Quiescence is an energy saving state, characterized by the acquisition of a catabolic status. In quiescent HSCs there is a reduction in energy consumption, protein synthesis and ROS accumulation. Meanwhile, HSC expansion and differentiation requires an anabolic state where increased energy consumption and anabolic processes such protein synthesis gain dominance (Tadokoro & Hirao, 2022).

2.2.1 FAOtastic beasts and where to find them

A great amount of energy is expended daily by the BM in an attempt to sustain hematopoietic turnover and keep all of its regulatory networks in check (Confavreux, Levine, & Karsenty, 2009; Rendina-Ruedy & Rosen, 2020). While glucose and its associated catabolism are widely used by all marrow cells during states of high energetic demand such as anemia or sepsis, another valuable source of energy lies within the adipocytes in the BM: lipids.

Lipids make up a very broad class of macromolecules used in a variety of cellular functions. Lipids make up the membranes of cells and organelles but are also a source of cellular energy and can also act as signaling molecules. In response to changes in physiological conditions, a cell's lipidome is actively remodeled, and imbalances in lipid homeostasis may contribute to downstream alterations in tissue homeostasis and regeneration (Clémot, Sênos Demarco, & Jones, 2020). The metabolic reactions that make up or break down lipids, have been shown to play an instructive role in stem cell fate decisions. In HSCs, the breakdown of lipids through PML-PPARδ-mediated fatty acid oxidation (FAO) has been shown to promote their asymmetric division and self-renewal (Ito et al., 2012). HSCs have higher rates of

FAO than their progeny, and FAO inhibition with etomoxir led to a significant reduction in LT-HSC occupancy upon transplantation. This suggests an important function for FAO downstream of PPARδ activation in enhancing HSC function. Interestingly, alongside increased mitophagy, Tie2+ HSCs also exhibited a marked activation of the PPAR-FAO axis. The high self-renewal capacity of these Tie2+ HSCs seemed to heavily rely on both; the maintenance of appropriate levels of mitophagy and the differential expression of FAO-related genes and a FAO metabolic signature (Ito et al., 2016).

With stem cell commitment comes an increase in anabolic demand. ACC, the rate-limiting enzyme in fatty acid synthesis (FAS), has been proposed to act as a differentiation rheostat by controlling the ratio of anabolic lipogenesis to catabolic FAO (Clémot et al., 2020; Folmes, Dzeja, Nelson, & Terzic, 2012; Folmes & Terzic, 2014; Ito et al., 2012). Overall, numerous HSC differentiation pathways seem to benefit from de novo lipid synthesis. Either through the role of lipids as signaling molecules, the way that the FAO/FAS ratio alters the energy/redox balance in the cell or just by allowing HSCs to expand their plasma and organelle membranes upon division. Cumulative evidence points at a vital role of lipids and their associated metabolites as regulators of a stem cell's transcriptome and epigenome (Clémot et al., 2020; Etchegaray & Mostoslavsky, 2016). A TNF-a driven increase in ceramide biosynthesis for example, has been shown to influence myelopoiesis through increased expression of key transcription factors like PU.1, GATA1 and GATA2 (Orsini et al., 2019; Pernes, Flynn, Lancaster, & Murphy, 2019). Similarly, the transcription factor SREBP-2 is also regulated by lipids (cholesterol), and its activation has been shown to be involved in HSC mobilization (Gu et al., 2019). Lipids can also significantly regulate gene expression by acting as substrates or co-factors of epigenetic modifiers. Proteomics and isotope tracing has revealed lipid metabolism as the main source of Acetyl-CoA in the cell. Acetyl-CoA is then used as substrate for histone acetylation and mediates the epigenetic activation of target genes (McDonnell et al., 2016). In HSCs, recent work by Umemoto and colleagues uncovered an important regulatory network linking Acetyl-CoA metabolism and the regulation of histone acetylation upon HSC expansion and differentiation. In their paper, the authors describe how HSCs respond to the needs of hematopoietic regeneration via dynamic changes in acetyl-CoA synthesis, its effect in H3K27ac levels and the downstream changes in epigenetic landscape (Umemoto et al., 2022). In their model, it is the correct timing in the increase and decrease of Acetyl-CoA metabolism and its associated epigenetic marks what allows HSCs to choose whether to differentiate or remain quiescent during regenerative hematopoiesis (Lisi-Vega & Méndez-Ferrer, 2022).

2.2.2 How the niche modulates lipid metabolism

For the longest time, BM adipocytes were thought to be just passive bystanders to all the crosstalk occurring in the BM. Even though they are the most abundant cell type in the BM, adipocytes and their associated lipid droplets were considered to act as "fillers" in space previously occupied by hematopoietic and skeletal elements in the BM (Rendina-Ruedy & Rosen, 2020). Now, these droplets are one of the defining features of adipocytes and are known to be incredibly dynamic. Their composition has even been used by some researchers to identify different BM adipocyte subtypes (Tratwal et al., 2021). Currently, lipid droplets have also been shown to accumulate in BMSCs and other more mature cells like macrophages. When energy is in demand, lipids inside these droplets can be catabolized by lipid-droplet-associated lipases or lipophagy. The process of lipolysis allows BM adipocytes to supply (at least in vitro) neighboring cells like OBs with fatty acid substrates (Maridas et al., 2019; Zhou, Yao, He, & Zhao, 2019). A factor to consider here is that fatty acids can have different downstream effects when taken up by neighboring cells. Palmitate seems to lead to an increase in mitochondrial ROS (Yuzefovych, Wilson, & Rachek, 2010), whereas oleate uptake has been shown to counteract palmitate-induced mitochondrial damage (Gillet et al., 2015). Nevertheless, the role of adipocytes in HSC regulation and hematopoiesis is still unclear. Seeing how the FAO/FAS balance in HSCs can determine their division choice, the potential regulation of HSC lipid metabolism through lipids coming from the niche is still very much unexplored. Future studies should focus on confirming whether this type of metabolic crosstalk between HSCs and BM adipocytes occurs in vivo. So far, experimental evidence indicates that this may be the case, as HSC-derived cytokines have been shown to at least activate lipolysis on visceral adipose tissue (VAT) (Waite, Floyd, Arbour-Reily, & Stephens, 2000). Cytokines like IFN might exert their effects on HSCs through the activation of their cognate receptors but also through their effect on the metabolic crosstalk with the niche, favoring the release of free fatty acids by adipocytes and their effects on FAO/FAS levels in HSCs (Lee, Al-Sharea, Dragoljevic, & Murphy, 2018).

In situations of chronic or acute energy deficiency, adipocytes paradoxically tend to store more lipids. This lipid accumulation might reflect a protective mechanism whereby the BM niche tries to establish a reserve of readily available energy to maintain hematopoiesis during nutrient stress. This could explain the homing of immune cells to the BM and their switch to lipid metabolism and supports the possibility of the BM acting as a metabolically "safe harbor."

The role of BMSCs on BM niche lipid metabolism is even less characterized. Most of the studies done to date look at the changes occurring in the lipid profile of MSCs after in vitro culture, with the main aim of identifying lipid changes that might become relevant when it comes to harnessing the therapeutic potential in the clinic (Chatgilialoglu et al., 2017; Lu et al., 2019). This limits the conclusions that one can draw from these experiments and makes it difficult to interpret in the context of HSC-niche metabolic crosstalk. One consistent observation is the dynamic nature of the BMSC lipidome, which seems to change during differentiation and under inflammatory stimuli (Campos et al., 2016; Clémot et al., 2020). This indicates a potential role for lipids in the immunomodulatory properties of BMSCs. It would be interesting to see how these variations in BMSC lipids (and perhaps lipid metabolism) come into play in diseases with very marked inflammatory profiles such as MPNs, MDS, etc. Overall, there is little evidence on the fate of these BMSC-derived lipids within the BM niche. Some groups have reported that gangliosides like GM1, GM331 or GD1a can be incorporated by myeloid-like cells and support their differentiation (Pernes et al., 2019; Santos et al., 2011; Ziulkoski et al., 2006). This evidence backs up a possible lipid-mediated metabolic crosstalk between BMSCs and progenitors, which might support myelopoiesis. In fact, with age, the lipid content of the BM increases, the central BM niche expands (Ho et al., 2019), and myeloid skewing can be observed in the hematopoietic compartment.

2.2.3 Protein synthesis and amino acid metabolism in HSC maintenance and differentiation

Translation is one of the most energy costly processes within the cell, with any proliferating eukaryotic cell being estimated to expend up to 80% of its energy in ribosome biogenesis (Schmidt, 1999). The metabolically demanding task of maintaining tight regulated translation comes however, with tremendous benefit. This is because alterations in translational control precede any other changes affecting gene expression and therefore represent powerful means of adaptation to any stressors and evolutionary pressures that a cell may encounter across its lifespan (Bhat et al., 2015). It has been calculated that at any given timepoint a single mammalian cell accommodates, on average, a cytoplasmic pool of 10^5 to 10^6 ribosomes (Bastide & David, 2018; Gupta & Warner, 2014). It is the availability of "free ribosomes" within this cytoplasmic pool what is considered the limiting factor governing protein synthesis rates. To fully reap the benefits of an increase in anabolic output, a cell must increase this "free ribosome" pool to convert these newly available macromolecules into functional proteins. The cellular pathways regulating metabolism and protein translation are therefore strongly coupled.

Different lines of evidence underscore the importance of tightly regulated translation in the hematopoietic system. First of all, hematological malignancies are among the cancers that most prominently benefit from increased protein synthesis to foster their development and progression (Ruggero & Pandolfi, 2003). Secondly, human disorders characterized by ribosomal dysfunction or defects in ribosome biogenesis possess one common hallmark—impaired hematopoiesis (Sulima & Keersmaecker, 2018). In HSCs and progenitors, protein synthesis seems to follow a pattern that complements their metabolic profile. A study by Signer and colleagues determined that HSCs synthesize less protein than other hematopoietic progenitors and that this difference in protein synthesis seems to be due to increased translational inhibition via 4E-BP proteins (Signer, Magee, Salic, & Morrison, 2014). Deletion of these 4E-BP proteins alone already increased protein synthesis and at the same time impaired engraftment of HSCs in serial transplantation experiments (Signer et al., 2016). This is in accordance with previous observations where HSC quiescence very much equates into a catabolic rather than anabolic profile. Low levels of protein synthesis also mean that HSCs are at lower risk of accumulating misfolded proteins, which are difficult to clear in rarely dividing cells (Schüler, Gebert, & Ori, 2020). A recent study determined that although LSKs exhibit low global protein synthesis rates, their translational program is characterized by high translational efficiencies of mRNAs involved in HSC maintenance (Spevak et al., 2020). Conversely, the authors revealed that a different translational program exists in myeloid progenitors (MPs), where their higher global translation is sustained by an mTOR-independent mechanism via CDK1 phosphorylation of eIF4E's negative regulators (4E-BP). Aberrant expression of mTOR in MPs resulted in an increased number of myeloid cells, underscoring how metabolic changes during lineage commitment result in anabolic pathways driving high levels of protein synthesis and conducing to HSC differentiation.

2.2.4 *The role of the BM niche and its metabolites in HSC protein synthesis*

From a metabolic standpoint, amino acids are the most important metabolite a cell needs to successfully translate mRNA into its protein products. In conditions of nutrient scarcity and low amino acid levels, tRNAs cannot be charged with their corresponding amino acid and mRNA translation comes to a halt (Advani & Ivanov, 2019). This often triggers a cellular stress response like the integrated stress response (ISR), which diverts the output of the cell's anabolic pathways with important implications for HSC function (van Galen et al., 2018).

The control of the amino acid supply and HSC amino acid biosynthetic pathways allows the BM niche to indirectly regulate HSC translation. For example, we know that HSCs are highly sensitive to changes in the abundance of certain amino acids like valine (Ito et al., 2019; Taya et al., 2016) and that imbalances in branched chain amino acid levels within the niche affect HSC survival and proliferation (Wilkinson, Morita, Nakauchia, & Yamazaki, 2018). To date, any crosstalk involving HSCs and translation-related metabolites has been described in the context of physiological stress. David Scadden's group recently reported the extracellular vesicle (EV) mediated transfer of tiRNAs from osteoblasts to GMPs under genotoxic or infectious stressors. This tiRNA/tRNA cargo was capable of increasing protein translation in the receiving GMPs, leading to cell proliferation and myeloid differentiation (Kfoury et al., 2021). Whether this metabolic crosstalk between HSCs and the niche occurs in homeostatic conditions and has downstream effects on HSC protein synthesis rates and differentiation remains to be investigated (Fig. 1).

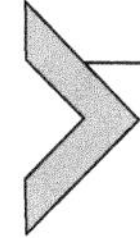

3. The metabolic reprogramming of the BM stem-cell niche in disease

The intricate structure of the BM, along with its various cell types and signaling networks, has evolved to function as a gatekeeper of hematopoiesis (Fröbel et al., 2021). The interdependence between hematopoiesis and the niche is a crucial factor in disease. Disruption of any signals or constituents within the niche is likely to significantly impact the others, making it challenging to unravel these complex interactions. In this section, we will provide a brief discussion of the most notable findings related to the metabolic crosstalk within the BM niche during aging, metabolic disease, and hematological malignancy.

The pivotal role of the niche in controlling hematopoiesis has motivated researchers to explore its deregulation as either a cause or consequence of disease. At a macroscopic level, substantial differences can already be observed in the BM architecture of older individuals and patients with hematological malignancies. For instance, during aging, the central niche expands while the endosteal niche decreases (Ho et al., 2019). Conversely, BM fibrosis and alterations in blood vessel density are common characteristics of various hematological malignancies, such as lymphoid and myeloid neoplasms (Gleitz, Kramann, & Schneider, 2018).

Beyond the macroscopic level, we now understand that both cell-intrinsic and niche-derived metabolic intermediates and signals play a role in regulating stem cell behavior and, consequently, tissue maintenance. The BM niche and its metabolic networks are altered in disease. This raises the question of whether alterations in metabolites within the niche and the stem cells themselves can contribute to disease progression. For example, changes in HSC metabolism are sometimes associated with a premature aging phenotype. Cell-intrinsic alterations in HSCs such as mutations affecting key enzymes in the tricarboxylic acid (TCA) cycle, such as IDH1/2, as well as the production of the oncometabolite 2-hydroxyglutarate (2-HG), have been shown to lead to abnormal self-renewal and malignant transformation (DiNardo et al., 2013; Issa & Dinardo, 2021; Mardis et al., 2009).

3.1 Metabolic crosstalk in aging

Aging and hematological malignancies are characterized by niche and hematopoietic dysfunction, and significant remodeling of the BM cavity. Reactive oxygen species (ROS) are one of the main drivers of aging and hematopoietic cell dysfunction within the BM. ROS is produced endogenously as a by-product of normal cellular metabolism. At physiological levels, ROS acts as a signaling molecule, regulating various cellular processes, including HSC fate (Ludin et al., 2014). However, when the brilliant metabolic regulation of HSCs is compromised and ROS levels exceed the cell's antioxidant capacity, oxidative stress occurs (Huang et al., 2019). Excessive oxidative stress can result in lipid and nucleic acid oxidation, cell apoptosis, senescence, and malignant cell transformation (Finkel & Holbrook, 2000).

HSCs are more vulnerable to oxidative stress than their progeny. High levels of ROS can cause loss of stemness and even cell death in HSCs. Studies using a ROS indicator (DCFDA) have shown that HSCs with low ROS levels contribute a higher donor repopulation percentage than HSCs with high ROS levels (Jang & Sharkis, 2007). Excessive mitochondria or elevated oxidative phosphorylation (OXPHOS) levels can increase ROS production, triggering oxidative stress and ultimately leading to loss of self-renewal capacity. This emphasizes how HSC metabolic output must be tightly regulated to maintain their quiescent state and self-renewal capacity. Apart from adapting their metabolic profile to keep ROS levels in check, HSCs can employ other strategies to downregulate ROS. For example, by activating specific transcriptional programs driven by polycomb complex Bmi1 (Park et al., 2003), or the transcription factor FOXO3a (Miyamoto et al., 2007; Tothova et al., 2007). Low anabolism and a reliance on glycolysis enable HSCs to maintain low ROS levels and quiescence and explains their distinct metabolic profile. The involvement of anabolic pathways like mTORC1 in HSC aging, through the regulation of proteins like Wip1, further supports this notion (Chen, Yi, et al., 2015). Under homeostatic conditions, HSCs at the top of the hierarchy maintain a high self-renewal capacity by efficiently disposing of damaged mitochondria. In aged mice, approximately one-third of HSCs retain a regenerative potential similar to that of young HSCs. These metabolically unique HSCs exhibit a low metabolic state coupled with high levels of autophagy. Symmetrical division poses a risk of accumulating damaged organelles in quiescent daughter cells, making cellular processes like autophagy and mitophagy crucial for maintaining HSC self-renewal (Ito et al., 2019). Conditional deletion of Atg12, a gene involved in autophagy, in HSCs has been shown to result in myeloid bias, reduced regenerative potential, and epigenetic alterations, along with elevated OXPHOS levels (Ho et al., 2017). These findings highlight the widespread effects that metabolic dysregulation can have on the HSC aging phenotype. Changes in the metabolic crosstalk between HSCs and the BM niche can profoundly impact these aging-related cellular processes. For instance, BMSCs may transfer an excessive number of mitochondria to HSCs in response to infection or regenerative hematopoiesis, leading to elevated OXPHOS levels and ROS production which might prompt HSCs to exit quiescence. Higher than normal OXPHOS levels or an overall more active metabolic state (through the stimulation of glucose or lipid metabolism) can induce oxidative stress in HSCs and have an undesired effect on those metabolic pathways and transcriptional programs regulating HSC self-renewal. Metabolic changes are therefore closely implicated in the age-related decline of stem cell function (Clémot et al., 2020; Ren, Ocampo, Liu, Carlos, & Belmonte, 2017).

Loss of lipid homeostasis and alterations in tissue-specific lipid profiles have been observed with age. In the BM, increased fat storage affects membrane lipid composition and decreases membrane fluidity (Gonzalez-Covarrubias, 2013). In turn, these changes in membrane fluidity are likely influence lipid raft formation. Lipid rafts are dynamic membrane microdomains that serve as platforms for signal transduction and membrane trafficking (Clémot et al., 2020; Simons & Ehehalt, 2002). Mouse HSCs are enriched for lipid rafts compared to progenitors, and the clustering of these structures is essential for the regulation of various signaling pathways involved in HSC quiescence, such as PI3K-AKT-FOXO or JAK/STAT (Hermetet et al., 2019; Yamazaki et al., 2006). However, how exactly alterations in BM lipid homeostasis influence the metabolic profile of HSCs and contribute to age-related changes in HSC function remains largely unexplored.

Another noticeable anatomical change during aging in the BM is the organization of blood vessels. Certain regions, such as the transition zone, experience a decline in blood vessel density, while others, like the central marrow, are relatively unaffected (Y. H. Ho et al., 2019; Y. H. Ho & Méndez-Ferrer, 2020). This can result in changes in nutrient supply and alterations to the existing oxygen gradient, significantly influencing HSC behavior through direct effects on OXPHOS and glycolysis, or indirect activation of Hif-1a. Interestingly, perisinusoidal niches maintain their shape and morphology to a greater extent with age, and HSCs located close to these niches exhibit preserved functionality (Saçma et al., 2019). It is conceivable that different locations within the BM niche have modulatory effects on HSC metabolism, thus playing a significant role during aging in the ability of HSCs to maintain a quiescent metabolic profile that preserves their self-renewal capacity (Schüler et al., 2020).

3.2 Metabolic crosstalk within metabolic disorders

3.2.1 Obesity

Overt changes in the function of HSCs and their surrounding niche are not exclusive to aging but also occur in metabolic disorders such as diabetes or obesity. These disorders have profound effects on systemic metabolism and as a result, also impact the metabolic features of the BM. As mentioned earlier, lipid rafts play a crucial role in integrating some of the signals involved in HSC quiescence. When mice are fed a high-fat diet (HFD), their HSCs exhibit enlarged lipid rafts, leading to altered distribution of their membrane receptors. This can decrease signaling of some receptors like

TGFβ and result in abnormal re-entry of HSCs into the cell cycle, ultimately causing HSC exhaustion (Clémot et al., 2020). Prolonged exposure to HFD has been shown to increase HSC numbers and skew differentiation towards the myeloid lineage, which further contributes to the persistent inflammation found in metabolic disease (Singer et al., 2014). Notably, the effects of obesity on systemic lipid availability, the BM niche, and HSC function can even be observed in the short term. Short-term exposure to HFD promotes HSC proliferation and enhances myeloid differentiation potential, while long-term HSCs (LT-HSCs) decrease in number (Hermetet et al., 2019; Luo et al., 2015; Van Den Berg et al., 2016).

These observed effects may be attributed to the influence of increased lipid availability on the metabolic profile of HSCs, such as variations in the FAO/FAS ratio and their impact on HSC fate. However, some authors propose a specific cell-intrinsic effect of HFD on HSCs (Hermetet et al., 2019), while others suggest that the main effects of short-term HFD are primarily cell-extrinsic, mediated by the impact of HFD on BMSC differentiation and changes in PPARγ activation (Luo et al., 2015). The cell-intrinsic nature of certain obesity-mediated alterations is further illustrated by their persistence upon serial BM transplantation. Transplanted HSPCs from obese mice maintain their preference for myeloid expansion and give rise to inflammatory adipose tissue macrophages (Amano et al., 2014; Singer et al., 2014; Xu et al., 2022). However, it is also evident that cell-extrinsic mechanisms contribute to the metabolic changes observed in obesity. HFD consumption disrupts the architecture of the bone and BM, promoting the differentiation of BMSCs into adipocytes and impairing osteoblastogenesis, causing a reduction in bone mass and impacting BM niche organization. Furthermore, HSCs mobilize to the spleen in obesity, indicating a clear microenvironmental influence. Along these lines, while the serial transplantation of HSPCs recapitulates some of the phenotype, so does the transplantation of visceral adipose tissue (VAT). When transplanted into WT mice, VAT from leptin-deficient Ob/Ob mice was capable of recapitulating the myeloid skewing, neutrophilia, and monocytosis characteristic of obesity (Nagareddy et al., 2014). The extent to which altered metabolic crosstalk mediates this BM phenotype is demonstrated by the partial reversal of obesity-related changes through physical activity and dietary intervention. Exercise has been shown to increase the production of HSPC quiescence factors by the niche (Frodermann et al., 2019), while caloric restriction improves bone density and immunological memory (Collins et al., 2019; Gerbaix et al., 2013). Based on the available data, it can be concluded that these two hypotheses are not mutually exclusive, and that the effects observed in obesity are likely to result from a combination of various cell-intrinsic and extrinsic mechanisms.

3.2.2 Diabetes

The effect of obesity on the hematopoietic system can be reproduced by exposing organisms to a high-fat diet (HFD), indicating that the increased lipid availability and altered lipid profile caused by HFD can be considered defining features of this metabolic disease. In the context of diabetes, it is crucial to examine the effects of hyperglycemia and dyslipidemia on HSC and BM niche function. We know that diabetic complications can interfere with hematopoietic function, often leading to enhanced myelopoiesis (Xu et al., 2022). Hence why it is important to understand the extent to which the diabetic phenotype can be attributed to changes in the metabolic crosstalk happening in the BM.

Persistent hyperglycemia in diabetes leads to an increase in proliferation (and number) of common myeloid progenitors (CMPs) and granulocyte-macrophage progenitors (GMPs), promoting myelopoiesis (Nagareddy et al., 2013). This increased production of myeloid cells is believed to contribute to the higher risk of cardiovascular disease (CVD) associated with diabetes. Partially targeting the altered metabolism by reducing glucose levels has been successful in reducing diabetes-related monocytosis. However, high glucose levels induce a form of memory in BM progenitors and a trained immune phenotype known as "hyperglycemic memory," which persists even after transplantation and may explain the high risk of CVD even after glucose-lowering therapy (Edgar et al., 2021; "The ACCORD Study Group", 2011). This continued risk of CVD likely arises from the interaction between the hyperglycemic and hypercholesterolemic phenotypes observed in diabetic patients.

On the other hand, altered lipid levels within the BM can also have a significant effect on HSC quiescence and proliferation. In a type 1 diabetes model, cholesterol transporter expression is suppressed in GMPs and CMPs, resulting in intracellular cholesterol accumulation in these progenitors, particularly in lipid rafts. This ultimately leads to an enhanced proliferative response in HSPCs of diabetic patients. Hypercholesterolemia has also been reported to increase reactive oxygen species (ROS) generated by HSPCs, promoting loss of quiescence and expansion. The effects of hypercholesterolemia on HSPC homeostasis have been directly linked to the regulation of specific cell-intrinsic targets or pathways such as p38 MAPK or pRb (Ito et al., 2006; Seijkens et al., 2014; Tie, Messina, Yan, Messina, & Messina, 2014). However, its effects could also be attributed to the influence of high lipid levels on the metabolic status of HSPCs and their crosstalk with other components of the BM. This hypothesis is supported by the observed stem cell mobilization and extramedullary hematopoiesis reported in diabetic mice (Robbins et al., 2012; Swirski et al., 2007).

3.2.3 Concluding remarks on metabolic disorders

Despite their different etiology, most metabolic disorders, share commonalities in terms of the changes they induce in the BM (e.g., diabetes vs. obesity vs. anorexia). Whether there is an excess of fatty acids, glucose, or a lack of nutrients, one of the primary responses of the BM niche is to promote BM adipogenesis at the expense of bone formation. This suggests that alterations in the metabolic properties of the niche may represent a general response of the BM to systemic stress. To gain a comprehensive understanding of the mechanisms underlying hematopoietic disruption in metabolic disorders, it is necessary to go beyond an "inflammation-centric" perspective and examine metabolism-driven changes in HSPSCs, BMSCs and other components of the BM. By considering the metabolic changes occurring in both the BM niche and the hematopoietic cells themselves, researchers can uncover the complex interplay between systemic metabolism and hematopoiesis. Unraveling these intricate mechanisms will contribute to a deeper understanding of the pathophysiology of metabolic disorders and facilitate the development of targeted therapeutic interventions aimed at restoring hematopoietic homeostasis in these conditions.

3.3 The metabolic "rewiring" of the BM niche in malignancy

The changes that occur in the BM niche during malignancy have been extensively studied. As a general rule, malignant hematopoietic cells exhibit a distinct metabolic profile and exploit the metabolic reprogramming of the BM niche to promote their expansion at the expense of healthy HSCs. During malignancy, malignant cells colonize and progressively remodel HSC niches, inducing profound changes that lead to a complete functional "rewiring" of the BM. Leukemic stem cells but also metastatic breast cancer cells can be found in perivascular niches where they take advantage of resident BMSCs and ECs, and their secreted molecules such as CXCL12 or TGFβ (Nobre et al., 2021; Puneet Agarwal, Isringhausen, & Li, 2019; Tirado et al., 2023; Viñado et al., 2022). These cancer cells end up developing a high level of heterogeneity which is reflected by their different transcriptomic epigenetic, and most importantly for this review, metabolic profiles (Erdem et al., 2022; Klco et al., 2014; Li et al., 2020). The proliferation of these distinct clones around the different niches in the BM eventually leads to profound structural changes which severely impact normal hematopoiesis.

The BM niche can be seen as a facilitator of malignant transformation, enabling the expansion of pre-existing leukemic clones. Alternatively, it can be viewed as a direct driver of malignant hematopoiesis. In fact, niche-related stressors, such as reactive oxygen species (ROS) or the acquisition of niche-specific mutations, play a crucial role in the development of full-blown malignancy. For example, an inflammatory microenvironment is thought to promote the expansion of neoplastic clones. During malignancy, components of the microenvironment, such as BMSCs, support and protect these clones through a variety of mechanisms, including neoangiogenesis, activation of survival pathways, protection from oxidative stress, and immunosuppression (Méndez-Ferrer et al., 2020). The extensive remodeling of the niche can even determine patient symptoms and disease progression, as seen in BM fibrosis. While efforts have focused on understanding the interactions between malignant clones and niche-derived cytokines, growth factors, and adhesion molecules, it is essential to investigate the metabolic context of the changes seen in the niche (Solimando et al., 2022; Zahr et al., 2016). Understanding the metabolic crosstalk between malignant cells and the BM niche will provide valuable insights into disease progression and is likely to lead to the development of targeted therapeutic strategies.

3.3.1 The BM niche changes during malignancy

The reprogramming of the BM niche plays a significant role in facilitating the proliferation of malignant clones, protecting them from stressors such as chemotherapy, and promoting the recurrence of leukemia after treatment. Exploiting the crosstalk between malignant cells and the BM niche represents a promising therapeutic approach to prevent disease progression and relapse. Cancer metabolism is characterized by a complex network of pathways and unique metabolic signatures. Malignant cells heavily rely on these pathways to sustain their continuous growth, becoming dependent on high nutrient consumption rates to the point of developing remarkable metabolic plasticity just to meet their energy demands (Xu et al., 2021). The signals and pathways involved in the metabolic interplay between malignant cells and the niche are therefore one of the critical vulnerabilities yet to be targeted in hematological malignancies. In this section, we will use acute myeloid leukemia (AML), the most common type of leukemia in adults, to exemplify how metabolic crosstalk is altered in malignancy.

In AML, malignant cells progressively remodel the endosteal niche, leading to the loss of endosteal vessels, osteoblasts (OBs), and sympathetic nerves (Baryawno et al., 2019; Duarte et al., 2018; Hanoun et al., 2014). Imaging studies using a NADH/NAD+ probe have shown that glycolytic AML leukemic stem cells (LSCs) preferentially home to the more hypoxic endosteal niche rather than the central marrow (Hao et al., 2019). Within this niche, AML LSCs can maintain their symmetric cell divisions through the regulation of the glycolytic enzyme PDK2. The endosteal niche also exhibits higher ATP levels compared to the central niche, which may contribute to the preference of AML LSCs for this location (He et al., 2021). Endosteal-derived ATP is proposed to act through P2X7, enhancing Ca2+-mediated phosphorylation of CREB and inducing phosphoglycerate dehydrogenase (PHGDH) expression which enables LSCs to maintain serine metabolism, another critical determinant of LSC fate. ATP level variations across different niches in the BM (endosteal vs. central) likely correlate with the metabolic status of surrounding BM components such as OBs, OCs, BMSCs or ECs. Understanding the spatial relationships between metabolically distinct subsets of these cell types will help determine the existence of different metabolic niches within the BM and whether these might dictate the physical location of malignant cells based on their metabolic preferences.

Bone remodeling is also an energetically demanding process occurring in the endosteal niche, which can also influence the behavior of malignant clones. Under certain conditions such as G-CSF stimulation, expansion of malignant clones seems to be spatially restricted to BM cavities undergoing active bone remodeling (Christodoulou et al., 2020). Interestingly, in AML with the MLL-AF9 fusion, endosteal OBs, endothelium, and stroma are degraded while the central marrow and its vasculature are preserved (Duarte et al., 2018). The impact on the endosteal niche and bone metabolism also extends to other malignancies like multiple myeloma (MM), where patients exhibit skeletal alterations such as osteolytic bone destruction, osteoclastic bone resorption, and hypercalcemia. These observations underscore the significance of bone and its turnover in the maintenance of malignant cells (H. Zhang et al., 2023).

The reasons why certain malignant clones gain advantage over others and why disease progression occurs at varying rates and has different outcomes in patients with the same driver mutation remain unclear. The solution to this puzzle may lie in the coexistence of distinct metabolic and immunological environments within the BM. During malignancy, alterations in the bone and vasculature of the BM likely affect the metabolic networks responsible for coupling HSC location and metabolism with stem cell fate. The rewiring of these networks and the preferential homing of malignant cells to various niches within the BM would then contribute to metabolic heterogeneity and the subsequent selection of malignant clones. Ultimately, this could favor disease progression and relapse in some patients but not in others.

3.3.2 *Hypoxia and the rewiring of energy metabolism*

The metabolic characteristics of leukemia cells, play a crucial role when it comes to determining their survival, proliferation, and response to treatment. These cells exhibit altered metabolism compared to normal HSCs and rely on different metabolic pathways to meet their energetic and biosynthetic demands. The changes in the BM come hand in hand with changes in its metabolic characteristics. For example, the vascular leakiness frequently reported in malignancy increases hypoxia, promoting the growth of AML cells and protecting them from chemotherapy (Passaro et al., 2021; Tirado et al., 2023). The high proliferation rates of leukemic soon leads to a shortage in oxygen and nutrients, perpetuating a hypoxic, acidic and nutrient-poor environment within the BM (Xu et al., 2021). While this threatens the survival and maintenance of HSCs, the overactivation of Hif-1a is thought to facilitate the proliferation of malignant cells by promoting glycolysis, a metabolic pathway particularly exploited by LSCs. Remarkably, an upregulation in Hif-1a expression can be detected in the majority of AML patients (67%) and about half of MDS patients (49%), where it correlates with poor prognosis (Deeb et al., 2011; Tong, Hu, Zhuang, Wang, & Jin, 2012).

One key feature of leukemia metabolism is the preference for glycolysis, even in the presence of oxygen, known as the Warburg effect. More specifically, malignant cells exhibit aerobic glycolysis, converting glucose to lactate via fermentation of pyruvate, which subsequently enters the TCA cycle. Despite being less efficient than oxidative phosphorylation (OXPHOS), upregulation of glycolysis and the associated pentose phosphate pathway (PPP) enables malignant cells to sustain a higher rate of ATP production, providing ample energy and building blocks for cell division (B. Xu et al., 2021). The shift to aerobic glycolysis also offers the advantage of reducing harmful reactive oxygen species (ROS) and maintaining a balanced NAD+/NADPH ratio, potentially stabilizing Hif-1α, thereby creating a self-reinforcing positive feedback loop (Icard et al., 2018). Notably, elevated glucose content in AML patient blasts is associated with drug resistance and enhanced survival, mediated by the overexpression of glucose transporters and enzymes such as lactate dehydrogenase (Song et al., 2016). Furthermore, the deletion of AMP-activated protein kinase (AMPK) has been shown to suppress leukemogenesis through downregulation of the glucose transporter GLUT1, probably by impairing the uptake of glucose from the microenvironment (Saito, Chapple, Lin, Kitano, & Nakada, 2015).

Overall, leukemic stem cells (LSCs) and rapidly proliferating blasts are thought to have higher glucose content and glycolytic activity as a way of capitalizing on increased BM hypoxia and Hif-1α activation (Mesbahi, Trahair, Lock, & Connerty, 2022). However, studies in AML have reported both pro-tumorigenic and tumor-suppressive effects of Hif-1a, indicating that glycolytic metabolism in AML might not be solely under the direct control of Hif-1a and its regulators. This could potentially explain the discrepancies seen on the impact of Hif-1a activation on AML progression (Wierenga et al., 2019). These conflicting results could be attributed to the fact that despite utilizing glycolysis and its intermediates as a means of generating building blocks, AML cells, including LSCs, require mitochondrial metabolism to sustain their quiescence and self-renewal capacity (Mesbahi et al., 2022). Regardless of their low oxygen consumption and OXPHOS levels, LSCs still rely on this metabolic pathway (Erdem et al., 2022; Lagadinou et al., 2013; Tirado et al., 2023). In contrast, AML blasts exhibit high mitochondrial biogenesis and oxygen consumption (Škrtić et al., 2011). Interestingly, AML blasts possess increased mitochondrial numbers but display abnormal metabolic features such as low reserve capacity and poor ATP generation (Nelson et al., 2021; Sriskanthadevan et al., 2015). In the context of the BM niche, AML-derived ROS has been implicated in triggering the active transfer of functional mitochondria from BMSCs to AML cells through tunneling nanotubes (TNTs), promoting AML cell bioenergetics and antioxidant capacity, and in so doing, fueling leukemia cell growth and resistance to chemotherapy (Forte et al., 2020; Moschoi et al., 2016; Saito et al., 2021).

Mitochondria also play a pivotal role in coordinating and supporting other key metabolic processes such as the tricarboxylic acid (TCA) cycle or fatty acid oxidation (FAO). Notably, TCA cycle activity is upregulated in LSCs but not in AML blasts (Kuntz et al., 2017; Lagadinou et al., 2013). While glutamate serves as a viable alternative fuel source for AML blasts, LSCs primarily rely on fatty acids (FAs) to fuel their TCA cycle (German et al., 2016; Tcheng et al., 2021). This distinct metabolic profile sets LSCs apart from the bulk of the tumor and renders them particularly dependent on the supply of FAs derived from BM adipocytes in order to maintain this unique metabolic state (Tabe et al., 2017). AML cells ensure the availability of lipids in the niche by inducing lipolysis in BM adipocytes through the secretion of growth differentiation factor 15 (GDF-15), as well as promoting the release of long-chain FAs by upregulating fatty acid binding protein 4 (FABP4) (Lu et al., 2018; Shafat et al., 2017). In parallel to this, the overexpression of enzymes and fatty acid transporters such as CT2 or CD36 provides AML cells with an advantage in utilizing released FAs compared to hematopoietic stem cells (HSCs) (Stevens et al., 2020; B. Xu et al., 2021). AML cells can also prompt BMSCs to fuel their TCA cycle and lipid biosynthesis through the transfer of ROS via gap junctions. This process leads to the conversion of pyruvate to lactate in BMSCs and stimulates the secretion of acetate, which is subsequently consumed by AML cells (Vilaplana-Lopera et al., 2022). The significance of lipids derived from the microenvironment is underscored by the observation that after chemotherapy, gonadal adipose tissue harbors a significant enrichment of LSCs. These LSCs manage to survive treatment by harnessing the FAs generated through lipolysis to fulfill their metabolic requirements (Ye, Adane, Stranahan, Park, & Correspondence, 2016).

Cumulative data supports the idea that LSCs and AML blasts can be distinguished based on their distinctive metabolic properties. LSCs, characterized by a more "metabolically dormant" population, heavily rely on FAs to fuel the TCA cycle, while also depending on amino acid metabolism for OXPHOS-related energy consumption and survival (Mesbahi et al., 2022). In conditions of nutrient scarcity, AML cells utilize alternative carbon sources such as glutamine to meet their energy demands (Willems et al., 2013). Glutamate and glutamine can regulate OXPHOS levels by being converted to alpha-ketoglutarate and replenishing TCA cycle intermediates (Jacque et al., 2015). In fact, upregulation of glutaminolysis is a common metabolic adaptation observed during tyrosine kinase inhibitor treatment (Gallipoli et al., 2018; Gillet et al., 2015; Gregory et al., 2016). Additionally, BMSCs have been shown to transfer aspartate to AML blasts following chemotherapy, fueling the persistence of residual leukemia-initiating cells. Amino acid metabolism thus represents another feature that can discriminate LSCs. In stark contrast to HSCs, LSCs also highly express mTORC1, which allows LSCs to capitalize on the building blocks and energy produced but also makes them particularly vulnerable to its inhibition. As a matter of fact, main oncogenic signaling nodes, like mTOR and AMPK, are linked to the regulation of both protein synthesis and energy homeostasis, underscoring the importance of their co-regulation for leukemia cell survival and proliferation (Leibovitch & Topisirovic, 2018; Lindqvist, Tandoc, Topisirovic, & Furic, 2018). Overall, this sheds light on the crucial role of metabolic crosstalk between the niche and malignant cells, particularly when subjected to significant selective pressures such as chemotherapy.

3.3.3 Why we should target malignant metabolic crosstalk

Current treatment regimens demonstrate efficacy against the bulk of rapidly dividing leukemic cells, but they often fail to eliminate the LSC population, partly due to the protective nature of the BM microenvironment, which supports their quiescence and self-renewal capacity. In a similar fashion to HSCs, LSCs are maintained in a dormant state characterized by low levels of reactive oxygen species (ROS) and oxidative stress (Lagadinou et al., 2013; Mesbahi et al., 2022). These ROS-low LSCs also have decreased ATP and oxygen requirements, and rely on low-energy metabolic pathways. It is their distinctive metabolism, together with the metabolic support from the niche that allows LSCs to survive the extreme metabolic challenges posed by chemotherapy treatment (van Gastel et al., 2020). Chemotherapy treatment induces a transient state of stress in LSCs which drives leukemia regeneration through the emergence and selection of leukemia-regenerating cells, eventually leading to relapse (Farge et al., 2017). Notably, AML cells that persist after chemotherapy exhibit transient metabolic adaptations that enable their survival. For instance, LSCs from relapsed AML patients have been shown to resist venetoclax treatment by shifting their main energy source from amino acids to fatty acids (Jones et al., 2018).

Therefore, metabolic targeting is gaining momentum in hematological malignancies as a promising therapeutic strategy to selectively eliminate those malignant cells capable of resisting chemotherapy. In this context, the metabolic crosstalk with the BM niche plays a crucial role in leukemia cell adaptation by enhancing their bioenergetics and antioxidant capacity, and shuttling key metabolites such as aspartate to fuel their metabolic demands (Forte et al., 2020; van Gastel et al., 2020). Consequently, depleting specific metabolites crucial for LSC survival or suppressing metabolic pathways reliant on crosstalk with the niche represent promising therapeutic strategies to target chemoresistance. For example, drugs that stimulate BM adipogenesis (PPARγ agonists) have been shown to suppress AML self-renewal and restore healthy hematopoietic maturation (Boyd et al., 2017).

While metabolic reprogramming has been traditionally associated with oncogenic mutations and hypoxia, emerging perspectives suggest an unprecedented role of the immune microenvironment in driving the metabolic preferences of malignant cells (Tsai, Chuang, Lunt, Kaech, & Ho Correspondence, 2023). Recent research on melanoma has revealed that the tumor microenvironment can shape the metabolic reprogramming of cancer cells. In this model, tumor immunoediting guides tumor cells to preferentially engage specific metabolic programs that simultaneously support their proliferation and immune evasion. This raises intriguing questions; do similar tumor-immune interactions occur within the BM? How does tissue context influence this process? Or just generally, is the BM niche capable of steering the metabolic reprogramming of malignant cells? This knowledge could inform the development of innovative approaches that disrupt the metabolic support provided by the BM niche and enhance immune-mediated clearance of leukemic cells. Ultimately, research in this field holds the potential to improve treatment outcomes and enhance long-term remission rates in hematological malignancies.

3.4 Concluding remarks: How metabolic crosstalk can "fuel" the disease development and progression

The points discussed in this section really come to illustrate the need for a broader perspective that encompasses the various components of the BM niche, rather than solely focusing on cell-intrinsic mechanisms underlying disease. We ought to think of malignancy, aging, and metabolic disorders as diseases of the tissue. If Darwinian evolutionary principles are being used to explain the evolution of malignant clones, one also must consider that a cell's direct environment plays a critical role in determining evolutionary dynamics within a population (Pronk & Raaijmakers, 2019). The BM of patients with hematological malignancies or cancer predisposition syndromes, such as MDS, further proves the notion that these diseases are manifestations of tissue-level abnormalities rather than purely cell-autonomous disorders (Raaijmakers, 2012). As proposed in this review, metabolic crosstalk plays a more significant role than previously anticipated in shaping BM dynamics. This extends beyond homeostasis where the metabolic crosstalk with the niche can perpetuate, initiate, or reinforce disease mechanisms occurring within the BM (Fig. 2).

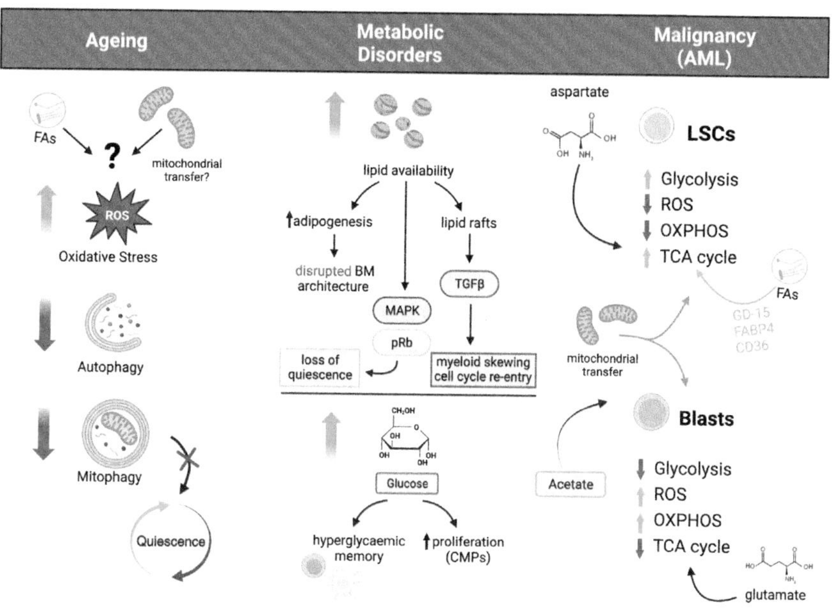

Fig. 2 A simplified view of the alterations in BM metabolic crosstalk during disease. Aging profoundly impacts the metabolic profile of HSCs and other hematopoietic cells (increases ROS, decreases autophagy and mitophagy levels), ultimately leading to the loss of quiescence and self-renewal capacity. Research has shown that the BM niche actively contributes to the aging phenotype, yet studies that interrogate the impact of the BM niche on the metabolic phenotype of aged HSCs are scarce. Metabolic disorders, such as obesity or diabetes, significantly affect the hematopoietic system. The rise in lipid and glucose levels leads to a change in the metabolic features of HSCs and the niche, resulting in increased HSC proliferation, loss of quiescence and myeloid skewing. These effects on hematopoietic output are believed to contribute to the persisting inflammation and disease complications (such as CVD) characteristic of these disorders. Lastly, during malignancy, a profound remodeling of the BM occurs, and the niche is "reprogrammed" to favor the expansion of malignant clones at the expense of healthy HSCs. From a therapeutic standpoint, it is crucial to consider that leukemic stem cells (LSCs) and the blasts comprising the tumor bulk exhibit different metabolic profiles. In this context, the metabolic support provided by the niche plays a pivotal role in promoting LSC survival following chemotherapy treatment. This occurs through a variety of mechanisms, like enhancing leukemia cell bioenergetics and antioxidant capacity, the shuttling of key metabolites or by supplying fatty acids to fuel the metabolic needs of residual leukemic cells after chemotherapy. Understanding these complex metabolic interactions within the BM niche holds significant promise for developing targeted therapeutic approaches to combat leukemia and improve treatment outcomes.

On this note, two critical aspects of this metabolic crosstalk have received limited attention in the existing literature: the concepts of metabolic coupling and metabolic heterogeneity within the niche. The terms metabolic asymmetry or "two-compartment metabolism" were coined to a phenomenon whereby the various cell types/compartments composing a tumor dramatically differ on their metabolic profiles (Guido et al., 2012; Valencia et al., 2014). This notion of metabolic coupling suggests that the BM, in order to fulfill its supportive function, is composed of niches that differ structurally and metabolically, i.e., is metabolically heterogeneous. It is imperative for future research to investigate the existence of metabolic heterogeneity within the BM niche and its potential role in disease progression. The field of hematology is only beginning to recognize the omnipresence of metabolism in all regulatory processes occurring within the BM. Further exploration of this metabolic crosstalk is likely to unveil a dynamic landscape of interactions that will help answer some of the outstanding questions in HSC and LSC biology.

References

ACCORD Study Group, Gerstein, H. C., Miller, M. E., Genuth, S., Ismail-Beigi, F., Buse, J. B., et al. (2011). Long-term effects of intensive glucose lowering on cardiovascular outcomes. *New England Journal of Medicine*, *364*(9), 818–828. https://doi.org/10.1056/nejmoa1006524/suppl_file/nejmoa1006524_disclosures.pdf.

Advani, V. M., & Ivanov, P. (2019). Translational control under stress: Reshaping the Translatome. *BioEssays*, *41*(5), 1900009. https://doi.org/10.1002/BIES.201900009.

Amano, S. U., Cohen, J. L., Vangala, P., Tencerova, M., Nicoloro, S. M., Yawe, J. C., et al. (2014). Local proliferation of macrophages contributes to obesity-associated adipose tissue inflammation. *Cell Metabolism*. https://doi.org/10.1016/j.cmet.2013.11.017.

Ambrosi, T. H., & Chan, C. K. F. (2021). Skeletal stem cells as the developmental origin of cellular niches for hematopoietic stem and progenitor cells. *Current Topics in Microbiology and Immunology*, *434*, 1–31. https://doi.org/10.1007/978-3-030-86016-5_1.

Ambrosi, T. H., Marecic, O., McArdle, A., Sinha, R., Gulati, G. S., Tong, X., et al. (2021). Aged skeletal stem cells generate an inflammatory degenerative niche. *Nature*, *256*, 597. https://doi.org/10.1038/s41586-021-03795-7.

Ambrosi, T. H., Scialdone, A., Graja, A., Sch€, A., Saraiva, L. R., Schulz Correspondence, T. J., et al. (2017). Adipocyte accumulation in the bone marrow during obesity and aging impairs stem cell-based hematopoietic and bone regeneration cell stem cell article adipocyte accumulation in the bone marrow during obesity and aging impairs stem cell-based hematopoietic and bone regeneration. *Cell Stem Cell*, *20*, 771–784. https://doi.org/10.1016/j.stem.2017.02.009.

Ansó, E., Weinberg, S. E., Diebold, L. P., Thompson, B. J., Malinge, S., Schumacker, P. T., et al. (2017). The mitochondrial respiratory chain is essential for haematopoietic stem cell function. *Nature Cell Biology*, *19*(6), 614–625. https://doi.org/10.1038/ncb3529.

Arai, F., Hirao, A., Ohmura, M., Sato, H., Matsuoka, S., Takubo, K., et al. (2004). Tie2/Angiopoietin-1 Signaling Regulates Hematopoietic Stem Cell Quiescence in the Bone Marrow Niche. *Cell*. https://doi.org/10.1016/j.cell.2004.07.004.

Asada, N., Kunisaki, Y., Pierce, H., Wang, Z., Fernandez, N. F., Birbrair, A., et al. (2017). Differential cytokine contributions of perivascular haematopoietic stem cell niches. *Nature Cell Biology, 19*(3), 214–223. https://doi.org/10.1038/ncb3475.

Baldridge, M. T., King, K. Y., Boles, N. C., Weksberg, D. C., & Goodell, M. A. (2010). Quiescent haematopoietic stem cells are activated by IFN-gamma in response to chronic infection. *Nature, 465*(7299), 793–797. https://doi.org/10.1038/NATURE09135.

Baryawno, N., Przybylski, D., Kowalczyk, M. S., Kfoury, Y., Severe, N., Gustafsson, K., et al. (2019). A cellular taxonomy of the bone marrow stroma in homeostasis and leukemia. *Cell, 177*(7), 1915–1932.e16. https://doi.org/10.1016/J.CELL.2019.04.040.

Bastide, A., & David, A. (2018). The ribosome, (slow) beating heart of cancer (stem) cell. *Oncogene*, 7(4), 1–13. https://doi.org/10.1038/s41389-018-0044-8.

Bhat, M., Robichaud, N., Hulea, L., Sonenberg, N., Pelletier, J., & Topisirovic, I. (2015). Targeting the translation machinery in cancer. In Nature reviews drug discovery (Vol. 14, Issue 4), pp. 261–278. Nature Publishing Group. https://doi.org/10.1038/nrd4505.

Bolamperti, S., Villa, I., & Rubinacci, A. (2022). Bone remodeling: An operational process ensuring survival and bone mechanical competence. *Bone Research.* https://doi.org/10.1038/s41413-022-00219-8.

Boyd, A. L., Reid, J. C., Salci, K. R., Aslostovar, L., Benoit, Y. D., Shapovalova, Z., et al. (2017). Acute myeloid leukaemia disrupts endogenous myelo-erythropoiesis by compromising the adipocyte bone marrow niche. *Nature Cell Biology, 19*(11), 1336–1347. https://doi.org/10.1038/NCB3625.

Bruns, I., Lucas, D., Pinho, S., Ahmed, J., Lambert, M. P., Kunisaki, Y., et al. (2014). Megakaryocytes regulate hematopoietic stem cell quiescence via Cxcl4 secretion. *Nature Medicine, 20*(11), 1315. https://doi.org/10.1038/NM.3707.

Buck, M. D., Sowell, R. T., Kaech, S. M., & Pearce, E. L. (2017). Metabolic instruction of immunity. *Cell, 169*(4), 570. https://doi.org/10.1016/J.CELL.2017.04.004.

Butler, J. M., Nolan, D. J., Vertes, E. L., Varnum-Finney, B., Kobayashi, H., Hooper, A. T., et al. (2010). Endothelial cells are essential for the self-renewal and repopulation of notch-dependent hematopoietic stem cells. *Cell Stem Cell, 6*(3), 251–264. https://doi.org/10.1016/J.STEM.2010.02.001.

Calvi, L. M., Adams, G. B., Weibrecht, K. W., Weber, J. M., Olson, D. P., Knight, M. C., et al. (2003). Osteoblastic cells regulate the haematopoietic stem cell niche. *Nature, 425*(6960), 841–846. https://doi.org/10.1038/NATURE02040.

Campos, A. M., Maciel, E., Moreira, A. S. P., Sousa, B., Melo, T., Domingues, P., et al. (2016). Lipidomics of mesenchymal stromal cells: Understanding the adaptation of phospholipid profile in response to pro-inflammatory cytokines. *Journal of Cellular Physiology, 231*(5), 1024–1032. https://doi.org/10.1002/JCP.25191.

Capitano, M. L., Mohamad, S. F., Cooper, S., Guo, B., Huang, X., Gunawan, A. M., et al. (2021). Mitigating oxygen stress enhances aged mouse hematopoietic stem cell numbers and function. *Journal of Clinical Investigation, 131*(1). https://doi.org/10.1172/JCI140177.

Chan, C. K. F., Chen, C. C., Luppen, C. A., Kim, J. B., DeBoer, A. T., Wei, K., et al. (2009). Endochondral ossification is required for haematopoietic stem-cell niche formation. *Nature, 457*(7228), 490–494. https://doi.org/10.1038/NATURE07547.

Chatgilialoglu, A., Rossi, M., Alviano, F., Poggi, P., Zannini, C., Marchionni, C., et al. (2017). Restored in vivo-like membrane lipidomics positively influence in vitro features of cultured mesenchymal stromal/stem cells derived from human placenta. *Stem Cell Research and Therapy, 8*(1), 1–11. https://doi.org/10.1186/S13287-017-0487-4/FIGURES/6.

Chen, Q., Liu, Y., Jeong, H. W., Stehling, M., Dinh, V. V., Zhou, B., et al. (2019). Apelin+ endothelial niche cells control hematopoiesis and mediate vascular regeneration after Myeloablative injury. *Cell Stem Cell*, *25*(6), 768–783.e6. https://doi.org/10.1016/J.STEM.2019.10.006.

Chen, X. L., Serrano, D., Mayhue, M., Hoebe, K., Ilangumaran, S., & Ramanathan, S. (2015). GIMAP5 deficiency is associated with increased AKT activity in T lymphocytes. *PLoS One*, *10*(10). https://doi.org/10.1371/JOURNAL.PONE.0139019.

Chen, Z., Yi, W., Morita, Y., Wang, H., Cong, Y., Liu, J. P., et al. (2015). Wip1 deficiency impairs haematopoietic stem cell function via p53 and mTORC1 pathways. *Nature Communications*, *6*(1), 1–11. https://doi.org/10.1038/ncomms7808.

Christodoulou, C., Spencer, J. A., Yeh, S. C. A., Turcotte, R., Kokkaliaris, K. D., Panero, R., et al. (2020). Live-animal imaging of native haematopoietic stem and progenitor cells. *Nature*, *578*(7794), 278–283. https://doi.org/10.1038/s41586-020-1971-z.

Clémot, M., Sênos Demarco, R., & Jones, D. L. (2020). Lipid mediated regulation of adult stem cell behavior. *Frontiers in Cell and Developmental Biology*, *8*, 115. https://doi.org/10.3389/FCELL.2020.00115/BIBTEX.

Collins, N., Han, S. J., Enamorado, M., Link, V. M., Huang, B., Moseman, E. A., et al. (2019). The bone marrow protects and optimizes immunological memory during dietary restriction. *Cell*, *178*(5), 1088–1101.e15. https://doi.org/10.1016/J.CELL.2019.07.049.

Confavreux, C. B., Levine, R. L., & Karsenty, G. (2009). A paradigm of integrative physiology, the crosstalk between bone and energy metabolisms. *Molecular and Cellular Endocrinology*, *310*(1–2), 21–29. https://doi.org/10.1016/J.MCE.2009.04.004.

Deeb, G., Vaughan, M. M., McInnis, I., Ford, L. A., Sait, S. N. J., Starostik, P., et al. (2011). Hypoxia-inducible factor-1α protein expression is associated with poor survival in normal karyotype adult acute myeloid leukemia. *Leukemia Research*, *35*(5), 579–584. https://doi.org/10.1016/J.LEUKRES.2010.10.020.

DiNardo, C. D., Propert, K. J., Loren, A. W., Paietta, E., Sun, Z., Levine, R. L., et al. (2013). Serum 2-hydroxyglutarate levels predict isocitrate dehydrogenase mutations and clinical outcome in acute myeloid leukemia. *Blood*, *121*(24), 4917. https://doi.org/10.1182/BLOOD-2013-03-493197.

Ding, L., Saunders, T. L., Enikolopov, G., & Morrison, S. J. (2012). Endothelial and perivascular cells maintain haematopoietic stem cells. *Nature*, *481*(7382), 457–462. https://doi.org/10.1038/nature10783.

Ding, Y., Song, N., & Luo, Y. (2012). Role of bone marrow-derived cells in angiogenesis: Focus on macrophages and pericytes. *Cancer Microenvironment: Official Journal of the International Cancer Microenvironment Society*, *5*(3), 225–236. https://doi.org/10.1007/S12307-012-0106-Y.

Dröscher, A. (2014). Images of cell trees, cell lines, and cell fates: The legacy of Ernst Haeckel and august Weismann in stem cell research. *History and Philosophy of the Life Sciences*, *36*(2), 157–186. https://doi.org/10.1007/S40656-014-0028-8/FIGURES/11.

Duarte, D., Hawkins, E. D., Akinduro, O., Ang, H., De Filippo, K., Kong, I. Y., et al. (2018). Inhibition of Endosteal vascular niche remodeling rescues hematopoietic stem cell loss in AML. *Cell Stem Cell*, *22*(1), 64–77.e6. https://doi.org/10.1016/J.STEM.2017.11.006.

Edgar, L., Akbar, N., Braithwaite, A. T., Krausgruber, T., Gallart-Ayala, H., Bailey, J., et al. (2021). Hyperglycemia induces trained immunity in macrophages and their precursors and promotes atherosclerosis. *Circulation*, *144*(12), 961–982. https://doi.org/10.1161/CIRCULATIONAHA.120.046464.

Eliasson, P., & Jönsson, J. I. (2010). The hematopoietic stem cell niche: Low in oxygen but a nice place to be. *Journal of Cellular Physiology*, *222*(1), 17–22. https://doi.org/10.1002/JCP.21908.

Ellis, S. L., & Nilsson, S. K. (2012). The location and cellular composition of the hemopoietic stem cell niche. *Cytotherapy*, *14*(2), 135–143. https://doi.org/10.3109/14653249.2011.630729.

Erdem, A., Marin, S., Pereira-Martins, D. A., Geugien, M., Cunningham, A., Pruis, M. G., et al. (2022). Inhibition of the succinyl dehydrogenase complex in acute myeloid leukemia leads to a lactate-fuelled respiratory metabolic vulnerability. *Nature Communications*, *13*(1). https://doi.org/10.1038/S41467-022-29639-0.

Etchegaray, J.-P., & Mostoslavsky, R. (2016). Molecular cell review interplay between metabolism and epigenetics: A nuclear adaptation to environmental changes. *Mol Cell*. https://doi.org/10.1016/j.molcel.2016.05.029.

Fadini, G. P., Ferraro, F., Quaini, F., Asahara, T., & Madeddu, P. (2014). Concise review: Diabetes, the bone marrow niche, and impaired vascular regeneration. *Stem Cells Translational Medicine*, *3*(8), 949–957. https://doi.org/10.5966/SCTM.2014-0052.

Fan, Y., Hanai, J., Le, P. T., Bi, R., Maridas, D., DeMambro, V., et al. (2017). Parathyroid hormone directs bone marrow mesenchymal cell fate. *Cell Metabolism*, *25*(3), 661–672. https://doi.org/10.1016/J.CMET.2017.01.001.

Farge, T., Saland, E., de Toni, F., Aroua, N., Hosseini, M., Perry, R., et al. (2017). Chemotherapy-resistant human acute myeloid leukemia cells are not enriched for leukemic stem cells but require oxidative metabolism. *Cancer Discovery*, 7(7), 716–735. https://doi.org/10.1158/2159-8290.CD-16-0441.

Fielding, C., García-García, A., Korn, C., Gadomski, S., Fang, Z., Reguera, J. L., et al. (2022). Cholinergic signals preserve haematopoietic stem cell quiescence during regenerative haematopoiesis. *Nature Communications*, *13*(1), 1–13. https://doi.org/10.1038/s41467-022-28175-1.

Filippi, M. D., & Ghaffari, S. (2019). Mitochondria in the maintenance of hematopoietic stem cells: New perspectives and opportunities. *Blood*, *133*(18), 1943. https://doi.org/10.1182/BLOOD-2018-10-808873.

Finkel, T., & Holbrook, N. J. (2000). Oxidants, oxidative stress and the biology of ageing. *Nature*. www.nature.com.

Folmes, C. D. L., Dzeja, P. P., Nelson, T. J., & Terzic, A. (2012). Metabolic plasticity in stem cell homeostasis and differentiation. *Cell Stem Cell*, *11*(5), 596–606. Cell Press https://doi.org/10.1016/j.stem.2012.10.002.

Folmes, C. D. L., & Terzic, A. (2014). Stem cell lineage specification: You become what you eat. *Cell Metabolism*, *20*(3), 389–391. Cell Press https://doi.org/10.1016/j.cmet.2014.08.006.

Forte, D., García-Fernández, M., Sánchez-Aguilera, A., Stavropoulou, V., Fielding, C., Martín-Pérez, D., et al. (2020). Bone marrow mesenchymal stem cells support acute myeloid leukemia bioenergetics and enhance antioxidant defense and escape from chemotherapy. *Cell Metabolism*, *32*(5), 829–843.e9. https://doi.org/10.1016/j.cmet.2020.09.001.

Fröbel, J., Landspersky, T., Percin, G., Schreck, C., Rahmig, S., Ori, A., et al. (2021). The hematopoietic bone marrow niche ecosystem. *Frontiers in Cell and Developmental Biology*, *9*. https://doi.org/10.3389/FCELL.2021.705410.

Frodermann, V., Rohde, D., Courties, G., Severe, N., Schloss, M. J., Amatullah, H., et al. (2019). Exercise reduces inflammatory cell production and cardiovascular inflammation via instruction of hematopoietic progenitor cells. *Nature Medicine*, *25*(11), 1761. https://doi.org/10.1038/S41591-019-0633-X.

Fujisaki, J., Wu, J., Carlson, A. L., Silberstein, L., Putheti, P., Larocca, R., et al. (2011). In vivo imaging of Treg cells providing immune privilege to the haematopoietic stem-cell niche. *Nature*, *474*(7350), 216–220. https://doi.org/10.1038/NATURE10160.

Galán-Díez, M., & Kousteni, S. (2017). The osteoblastic niche in hematopoiesis and hematological myeloid malignancies. *Current Molecular Biology Reports*, *3*(2), 53–62. https://doi.org/10.1007/S40610-017-0055-9.

Gallipoli, P., Giotopoulos, G., Tzelepis, K., Costa, A. S. H., Vohra, S., Medina-Perez, P., et al. (2018). Glutaminolysis is a metabolic dependency in FLT3 ITD acute myeloid leukemia unmasked by FLT3 tyrosine kinase inhibition. *Blood*, *131*(15), 1639–1653. https://doi.org/10.1182/blood-2017-12-820035.

Gan, B., Hu, J., Jiang, S., Liu, Y., Sahin, E., Zhuang, L., et al. (2010). Lkb1 regulates quiescence and metabolic homeostasis of haematopoietic stem cells. *Nature*, *468*(7324), 701–704. https://doi.org/10.1038/nature09595.

Gao, Q., Wang, L., Wang, S., Huang, B., Jing, Y., & Su, J. (2021). Bone marrow mesenchymal stromal cells: Identification, classification, and differentiation. *Frontiers in Cell and Developmental Biology*, *9*. https://doi.org/10.3389/FCELL.2021.787118.

García-García, A., Korn, C., García-Fernández, M., Domingues, O., Villadiego, J., Martín-Pérez, D., et al. (2019). Dual cholinergic signals regulate daily migration of hematopoietic stem cells and leukocytes. *Blood*, *133*(3), 224–236. https://doi.org/10.1182/BLOOD-2018-08-867648.

Gerbaix, M., Metz, L., Mac-Way, F., Lavet, C., Guillet, C., Walrand, S., et al. (2013). A well-balanced diet combined or not with exercise induces fat mass loss without any decrease of bone mass despite bone micro-architecture alterations in obese rat. *Original Full Length Article*. https://doi.org/10.1016/j.bone.2013.01.006.

German, N. J., Yoon, H., Yusuf, R. Z., Murphy, J. P., Finley, L. W. S., Laurent, G., et al. (2016). PHD3 loss in Cancer enables metabolic reliance on fatty acid oxidation via deactivation of ACC2. *Molecular Cell*, *63*(6), 1006–1020. https://doi.org/10.1016/J.MOLCEL.2016.08.014.

Ghazanfari, R., Li, H., Zacharaki, D., Lim, H. C., & Scheding, S. (2016). Human non-hematopoietic CD271pos/CD140alow/neg bone marrow stroma cells fulfill stringent stem cell criteria in serial transplantations. *Stem Cells and Development*, *25*(21), 1652–1658. https://doi.org/10.1089/scd.2016.0169.

Gillet, C., Spruyt, D., Rigutto, S., Dalla Valle, A., Berlier, J., Louis, C., et al. (2015). Oleate abrogates palmitate-induced lipotoxicity and Proinflammatory response in human bone marrow-derived mesenchymal stem cells and osteoblastic cells. *Endocrinology*, *156*(11), 4081–4093. https://doi.org/10.1210/EN.2015-1303.

Gleitz, H. F. E., Kramann, R., & Schneider, R. K. (2018). Understanding deregulated cellular and molecular dynamics in the haematopoietic stem cell niche to develop novel therapeutics for bone marrow fibrosis. *The Journal of Pathology*, *245*(2), 138. https://doi.org/10.1002/PATH.5078.

Golan, K., Singh, A. K., Kollet, O., Bertagna, M., Althoff, M. J., Khatib-Massalha, E., et al. (2020). Hematopoiesis and stem cells: Bone marrow regeneration requires mitochondrial transfer from donor Cx43-expressing hematopoietic progenitors to stroma. *Blood*, *136*(23), 2607. https://doi.org/10.1182/BLOOD.2020005399.

Goldberg, E. L., & Dixit, V. D. (2019). Bone marrow: An Immunometabolic refuge during energy depletion. *Cell Metabolism*, *30*(4), 621–623. https://doi.org/10.1016/j.cmet.2019.08.022.

Gonzalez-Covarrubias, V. (2013). Lipidomics in longevity and healthy aging. *Biogerontology*, *14*(6), 663–672. https://doi.org/10.1007/S10522-013-9450-7/TABLES/3.

Gregory, M. A., D'Alessandro, A., Alvarez-Calderon, F., Kim, J., Nemkov, T., Adane, B., et al. (2016). ATM/G6PD-driven redox metabolism promotes FLT3 inhibitor resistance in acute myeloid leukemia. *Proceedings of the National Academy of Sciences of the United States of America*, *113*(43), E6669–E6678. https://doi.org/10.1073/PNAS.1603876113/-/DCSUPPLEMENTAL/PNAS.1603876113.SD01.XLSX.

Gu, Q., Yang, X., Lv, J., Zhang, J., Xia, B., Kim, J., et al. (2019). AIBP-mediated cholesterol efflux instructs hematopoietic stem and progenitor cell fate. *Science (New York, N.Y.), 363*(6431), 1085. https://doi.org/10.1126/SCIENCE.AAV1749.

Guidi, N., Sacma, M., Ständker, L., Soller, K., Marka, G., Eiwen, K., et al. (2017). Osteopontin attenuates aging-associated phenotypes of hematopoietic stem cells. *The EMBO Journal, 36*(7), 840–853. https://doi.org/10.15252/EMBJ.201694969.

Guido, C., Whitaker-Menezes, D., Capparelli, C., Balliet, R., Lin, Z., Pestell, R. G., et al. (2012). Metabolic reprogramming of cancer-associated fibroblasts by TGF-β drives tumor growth: Connecting TGF-β signaling with "Warburg- like" cancer metabolism and L-lactate production. *Cell Cycle, 11*(16), 3019–3035. https://doi.org/10.4161/cc.21384.

Gupta, V., & Warner, J. R. (2014). Ribosome-omics of the human ribosome. *RNA, 20*(7), 1004–1013. Cold Spring Harbor Laboratory Press. https://doi.org/10.1261/rna.043653.113.

Hanoun, M., Zhang, D., Mizoguchi, T., Pinho, S., Pierce, H., Kunisaki, Y., et al. (2014). Acute myelogenous leukemia-induced sympathetic neuropathy promotes malignancy in an altered hematopoietic stem cell niche. *Cell Stem Cell, 15*(3), 365–375. https://doi.org/10.1016/J.STEM.2014.06.020.

Hao, X., Gu, H., Chen, C., Yu, Z., Yang, Y., & Correspondence, J. Z. (2019). Metabolic imaging reveals a unique preference of symmetric cell division and homing of leukemia-initiating cells in an Endosteal niche. *Cell Metabolism, 29*, 950–965. https://doi.org/10.1016/j.cmet.2018.11.013.

He, X., Wan, J., Yang, X., Zhang, X., Huang, D., Li, X., et al. (2021). Bone marrow niche ATP levels determine leukemia-initiating cell activity via P2X7 in leukemic models. *The Journal of Clinical Investigation, 131*(4). https://doi.org/10.1172/JCI140242.

Hermetet, F., Buffière, A., Aznague, A., Pais de Barros, J. P., Bastie, J. N., Delva, L., et al. (2019). High-fat diet disturbs lipid raft/TGF-β signaling-mediated maintenance of hematopoietic stem cells in mouse bone marrow. *Nature Communications, 10*(1). https://doi.org/10.1038/S41467-018-08228-0.

Ho, Y. H., del Toro, R., Rivera-Torres, J., Rak, J., Korn, C., García-García, A., et al. (2019). Remodeling of bone marrow hematopoietic stem cell niches promotes myeloid cell expansion during premature or physiological aging. *Cell Stem Cell, 25*(3), 407–418.e6. https://doi.org/10.1016/J.STEM.2019.06.007.

Ho, Y. H., & Méndez-Ferrer, S. (2020). Microenvironmental contributions to hematopoietic stem cell aging. *Haematologica, 105*(1), 38–46. https://doi.org/10.3324/HAEMATOL.2018.211334.

Ho, T. T., Warr, M. R., Adelman, E. R., Lansinger, O. M., Flach, J., Verovskaya, E. V., et al. (2017). Autophagy maintains the metabolism and function of young and old (hematopoietic) stem cells. *Nature, 543*(7644), 205. https://doi.org/10.1038/NATURE21388.

Hoyer, F. F., Zhang, X., Coppin, E., Vasamsetti, S. B., Modugu, G., Schloss, M. J., et al. (2020). Bone marrow endothelial cells regulate myelopoiesis in diabetes. *Circulation, 142*(3), 244. https://doi.org/10.1161/CIRCULATIONAHA.120.046038.

Hu, Q., Wu, D., Walker, M., Wang, P., Tian, R., & Wang, W. (2021). Genetically encoded biosensors for evaluating NAD+/NADH ratio in cytosolic and mitochondrial compartments. *Cell Reports Methods, 1*(7). https://doi.org/10.1016/J.CRMETH.2021.100116.

Huang, D., Chen, C., Xie, L., Yu, Z., & Zheng, J. (2019). Hematopoietic stem cell metabolism and stemness. *Blood Science, 1*(1), 12–18. https://doi.org/10.1097/BS9.0000000000000012.

Icard, P., Shulman, S., Farhat, D., Steyaert, J. M., Alifano, M., & Lincet, H. (2018). How the Warburg effect supports aggressiveness and drug resistance of cancer cells? *Drug Resistance Updates, 38*, 1–11. https://doi.org/10.1016/J.DRUP.2018.03.001.

Iommarini, L., Porcelli, A. M., Gasparre, G., & Kurelac, I. (2017). Non-canonical mechanisms regulating hypoxia-inducible factor 1 alpha in cancer. *Frontiers in Oncology*, 7, 322461. https://doi.org/10.3389/FONC.2017.00286/BIBTEX.

Issa, G. C., & Dinardo, C. D. (2021). Acute myeloid leukemia with IDH1 and IDH2 mutations: 2021 treatment algorithm. *Blood Cancer Journal*, *11*, 107. https://doi.org/10.1038/s41408-021-00497-1.

Itkin, T., Gur-Cohen, S., Spencer, J. A., Schajnovitz, A., Ramasamy, S. K., Kusumbe, A. P., et al. (2016). Distinct bone marrow blood vessels differentially regulate haematopoiesis. *Nature*, *532*(7599), 323–328. https://doi.org/10.1038/nature17624.

Ito, K., Bonora, M., & Ito, K. (2019). Metabolism as master of hematopoietic stem cell fate. *International Journal of Hematology*, *109*(1), 18. https://doi.org/10.1007/S12185-018-2534-Z.

Ito, K., Carracedo, A., Weiss, D., Arai, F., Ala, U., Avigan, D. E., et al. (2012). A PML-PPAR-δ pathway for fatty acid oxidation regulates hematopoietic stem cell maintenance. *Nature Medicine*, *18*(9), 1350–1358. https://doi.org/10.1038/nm.2882.

Ito, K., Hirao, A., Arai, F., Takubo, K., Matsuoka, S., Miyamoto, K., et al. (2006). Reactive oxygen species act through p38 MAPK to limit the lifespan of hematopoietic stem cells. *Nature Medicine*, *12*(4), 446–451. https://doi.org/10.1038/nm1388.

Ito, K., Turcotte, R., Cui, J., Zimmerman, S. E., Pinho, S., Mizoguchi, T., et al. (2016). Self-renewal of a purified Tie2+ hematopoietic stem cell population relies on mitochondrial clearance. *Science*, *354*(6316), 1156–1160. https://doi.org/10.1126/science.aaf5530/suppl_file/ito.sm.pdf.

Jacque, N., Ronchetti, A. M., Larrue, C., Meunier, G., Birsen, R., Willems, L., et al. (2015). Targeting glutaminolysis has antileukemic activity in acute myeloid leukemia and synergizes with BCL-2 inhibition. *Blood*, *126*(11), 1346–1356. https://doi.org/10.1182/BLOOD-2015-01-621870.

Jang, Y. Y., & Sharkis, S. J. (2007). A low level of reactive oxygen species selects for primitive hematopoietic stem cells that may reside in the low-oxygenic niche. *Blood*, *110*(8), 3056–3063. https://doi.org/10.1182/BLOOD-2007-05-087759.

Jin, G., Xu, C., Zhang, X., Long, J., Rezaeian, A. H., Liu, C., et al. (2018). Atad3a suppresses Pink1-dependent mitophagy to maintain homeostasis of hematopoietic progenitor cells. *Nature Immunology*, *19*(1), 29–40. https://doi.org/10.1038/S41590-017-0002-1.

Jones, C. L., Stevens, B. M., Alessandro, A. D.'., Degregori, J., Pollyea, D. A., & Correspondence, C. T. J. (2018). Inhibition of amino acid metabolism selectively targets human leukemia stem cells. *Cancer Cell*, *34*. https://doi.org/10.1016/j.ccell.2018.10.005.

Jordan, S., Tung, N., Casanova-Acebes, M., Chang, C., Cantoni, C., Zhang, D., et al. (2019). Dietary intake regulates the circulating inflammatory monocyte Pool. *Cell*, *178*(5), 1102–1114.e17. https://doi.org/10.1016/J.CELL.2019.07.050.

Kandarakov, O., Belyavsky, A., & Semenova, E. (2022). Bone marrow niches of hematopoietic stem and progenitor cells. *International Journal of Molecular Sciences*, *23*(8). https://doi.org/10.3390/IJMS23084462.

Kapp, F. G., Perlin, J. R., Hagedorn, E. J., Gansner, J. M., Schwarz, D. E., O'Connell, L. A., et al. (2018). Protection from UV light is an evolutionarily conserved feature of the haematopoietic niche. *Nature*, *558*(7710), 445–448. https://doi.org/10.1038/s41586-018-0213-0.

Kaushansky, K. (2006). Hematopoietic growth factors, signaling and the chronic myeloproliferative disorders. *Cytokine & Growth Factor Reviews*, *17*(6), 423–430. https://doi.org/10.1016/J.CYTOGFR.2006.09.005.

Kfoury, Y. S., Ji, F., Mazzola, M., Sykes, D. B., Scherer, A. K., Anselmo, A., et al. (2021). tiRNA signaling occurs via stress-regulated vesicle transfer in the hematopoietic niche. *Cell Stem Cell*, *28*(12), 2090–2103.e9. https://doi.org/10.1016/J.STEM.2021.08.014.

Kiel, M. J., Yilmaz, Ö. H., Iwashita, T., Yilmaz, O. H., Terhorst, C., & Morrison, S. J. (2005). SLAM family receptors distinguish hematopoietic stem and progenitor cells and reveal endothelial niches for stem cells. *Cell, 121*(7), 1109–1121. https://doi.org/10.1016/J.CELL.2005.05.026.

Kim, Y. W., Koo, B. K., Jeong, H. W., Yoon, M. J., Song, R., Shin, J., et al. (2008). Defective notch activation in microenvironment leads to myeloproliferative disease. *Blood, 112*(12), 4628–4638. https://doi.org/10.1182/BLOOD-2008-03-148999.

Kirkwood, T. B. L., Kapahi, P., & Shanley, D. P. (2000). Evolution, stress, and longevity. *Journal of Anatomy, 197*, 587–590.

Klco, J. M., Spencer, D. H., Miller, C. A., Griffith, M., Lamprecht, T. L., O'Laughlin, M., et al. (2014). Functional heterogeneity of genetically defined subclones in acute myeloid leukemia. *Cancer Cell, 25*(3), 379–392. https://doi.org/10.1016/J.CCR.2014.01.031.

Kocabas, F., Xie, L., Xie, J., Yu, Z., DeBerardinis, R. J., Kimura, W., et al. (2015). Hypoxic metabolism in human hematopoietic stem cells. *Cell & Bioscience, 5*(1). https://doi.org/10.1186/S13578-015-0020-3.

Kocabas, F., Zheng, J., Thet, S., Copeland, N. G., Jenkins, N. A., Deberardinis, R. J., et al. (2012). Meis1 regulates the metabolic phenotype and oxidant defense of hematopoietic stem cells. *Blood*. https://doi.org/10.1182/blood-2012.

Kode, A., Manavalan, J. S., Mosialou, I., Bhagat, G., Rathinam, C. V., Luo, N., et al. (2014). Leukaemogenesis induced by an activating β-catenin mutation in osteoblasts. *Nature, 506*(7487), 240–244. https://doi.org/10.1038/nature12883.

Kokkaliaris, K. D., & Scadden, D. T. (2020). Cell interactions in the bone marrow microenvironment affecting myeloid malignancies. *Blood Advances, 4*(15), 3795. https://doi.org/10.1182/bloodadvances.2020002127.

Koller, M. R., Bender, J. G., Terry Papoutsakis, E., & Miller, W. M. (1992). Beneficial effects of reduced oxygen tension and perfusion in Long-term hematopoietic cultures. *Ann N Y Acad Sci*. https://doi.org/10.1111/j.1749-6632.1992.tb42578.x.

Kollet, O., Dar, A., Shivtiel, S., Kalinkovich, A., Lapid, K., Sztainberg, Y., et al. (2006). Osteoclasts degrade endosteal components and promote mobilization of hematopoietic progenitor cells. *Nature Medicine, 12*(6). https://doi.org/10.1038/nm1417.

Kubota, Y., Takubo, K., & Suda, T. (2008). Bone marrow long label-retaining cells reside in the sinusoidal hypoxic niche. *Biochemical and Biophysical Research Communications, 366*(2), 335–339. https://doi.org/10.1016/J.BBRC.2007.11.086.

Kunisaki, Y., Bruns, I., Scheiermann, C., Ahmed, J., Pinho, S., Zhang, D., et al. (2013). Arteriolar niches maintain haematopoietic stem cell quiescence. *Nature, 502*(7473), 637–643. https://doi.org/10.1038/NATURE12612.

Kuntz, E. M., Baquero, P., Michie, A. M., Dunn, K., Tardito, S., Holyoake, T. L., et al. (2017). Targeting mitochondrial oxidative phosphorylation eradicates therapy-resistant chronic myeloid leukemia stem cells. *Nature Medicine, 23*(10), 1234–1240. https://doi.org/10.1038/nm.4399.

Lagadinou, E. D., Sach, A., Callahan, K., Rossi, R. M., Neering, S. J., Minhajuddin, M., et al. (2013). BCL-2 inhibition targets oxidative phosphorylation and selectively eradicates quiescent human leukemia stem cells. *Cell Stem Cell, 12*(3), 329–341. https://doi.org/10.1016/j.stem.2012.12.013.

Laurenti, E., & Göttgens, B. (2018). From haematopoietic stem cells to complex differentiation landscapes. *Nature*. https://doi.org/10.1038/nature25022.

Lee, M. K. S., Al-Sharea, A., Dragoljevic, D., & Murphy, A. J. (2018). Hand of FATe: Lipid metabolism in hematopoietic stem cells. *Current Opinion in Lipidology, 29*(3), 240–245. https://doi.org/10.1097/MOL.0000000000000500.

Leibovitch, M., & Topisirovic, I. (2018). Dysregulation of mRNA translation and energy metabolism in cancer. *Advances in Biological Regulation, 67*, 30–39. https://doi.org/10.1016/j.jbior.2017.11.001.

Lévesque, J. P., Helwani, F. M., & Winkler, I. G. (2010). The endosteal "osteoblastic" niche and its role in hematopoietic stem cell homing and mobilization. *Leukemia, 24*(12), 1979–1992. https://doi.org/10.1038/LEU.2010.214.

Lewis, J. W., Edwards, J. R., Naylor, A. J., & McGettrick, H. M. (2021). Adiponectin signalling in bone homeostasis, with age and in disease. *Bone Research, 9*(1), 1–11. https://doi.org/10.1038/s41413-020-00122-0.

Li, S., Chen, X., Wang, J., Meydan, C., Glass, J. L., Shih, A. H., et al. (2020). Somatic mutations drive specific, but reversible, epigenetic heterogeneity states in AML. *Cancer Discovery, 10*(12), 1934–1949. https://doi.org/10.1158/2159-8290.cd-19-0897/333426/am/somatic-mutations-drive-specific-but-reversible.

Li, W., Johnson, S. A., Shelley, W. C., & Yoder, M. C. (2004). Hematopoietic stem cell repopulating ability can be maintained in vitro by some primary endothelial cells. *Experimental Hematology, 32*(12), 1226–1237. https://doi.org/10.1016/j.exphem.2004.09.001.

Li, X., Sun, X., & Carmeliet, P. (2019). Hallmarks of endothelial cell metabolism in health and disease. *Cell Metabolism, 30*(3), 414–433. https://doi.org/10.1016/j.cmet.2019.08.011.

Lindqvist, L. M., Tandoc, K., Topisirovic, I., & Furic, L. (2018). Cross-talk between protein synthesis, energy metabolism and autophagy in cancer. *Current Opinion in Genetics and Development, 48*, 104–111. Elsevier Ltd. https://doi.org/10.1016/j.gde.2017.11.003.

Lisi-Vega, L. E., & Méndez-Ferrer, S. (2022). Stem cells "aclymatise" to regenerate the blood system. *The EMBO Journal, 41*(8). https://doi.org/10.15252/embj.2022110942.

Liu, X., Gu, Y., Kumar, S., Amin, S., Guo, Q., Wang, J., et al. (2023). Oxylipin-PPARγ-initiated adipocyte senescence propagates secondary senescence in the bone marrow. *Cell Metabolism, 35*(4), 667–684.e6. https://doi.org/10.1016/j.cmet.2023.03.005.

Lu, X., Chen, Y., Wang, H., Bai, Y., Zhao, J., Zhang, X., et al. (2019). Integrated Lipidomics and transcriptomics characterization upon aging-related changes of lipid species and pathways in human bone marrow mesenchymal stem cells. *Journal of Proteome Research, 18*(5), 2065–2077. https://doi.org/10.1021/ACS.JPROTEOME.8B00936/ASSET/IMAGES/LARGE/PR-2018-009366_0006.JPEG.

Lu, W., Wan, Y., Li, Z., Zhu, B., Yin, C., Liu, H., et al. (2018). Growth differentiation factor 15 contributes to marrow adipocyte remodeling in response to the growth of leukemic cells. *Journal of Experimental and Clinical Cancer Research, 37*(1), 1–10. https://doi.org/10.1186/s13046-018-0738-y/figures/5.

Luchsinger, L. L., Strikoudis, A., Danzl, N. M., Bush, E. C., Finlayson, M. O., Satwani, P., et al. (2019). Harnessing hematopoietic stem cell low intracellular calcium improves their maintenance in vitro. *Cell Stem Cell, 25*(2), 225–240.e7. https://doi.org/10.1016/j.stem.2019.05.002.

Ludin, A., Gur-Cohen, S., Golan, K., Kaufmann, K. B., Itkin, T., Medaglia, C., et al. (2014). Reactive oxygen species regulate hematopoietic stem cell self-renewal, migration and development, as well as their bone marrow microenvironment. *Antioxidants & Redox Signaling, 21*(11), 1605. https://doi.org/10.1089/ARS.2014.5941.

Luo, Y., Chen, G. L., Hannemann, N., Ipseiz, N., Krönke, G., Bäuerle, T., et al. (2015). Microbiota from obese mice regulate hematopoietic stem cell differentiation by altering the bone niche. *Cell Metabolism, 22*(5), 886–894. https://doi.org/10.1016/J.CMET.2015.08.020.

Małkiewicz, A., & Dziedzic, M. (2012). Bone marrow reconversion – Imaging of physiological changes in bone marrow. *Polish Journal of Radiology*, 77(4), 45. https://doi.org/10.12659/PJR.883628.

Manesia, J. K., Xu, Z., Broekaert, D., Boon, R., van Vliet, A., Eelen, G., et al. (2015). Highly proliferative primitive fetal liver hematopoietic stem cells are fueled by oxidative metabolic pathways. *Stem Cell Research, 15*(3), 715–721. https://doi.org/10.1016/J.SCR.2015.11.001.

Mardis, E. R., Ding, L., Dooling, D. J., Larson, D. E., McLellan, M. D., Chen, K., et al. (2009). Recurring mutations found by sequencing an acute myeloid leukemia genome. *The New England Journal of Medicine, 361*(11), 1058. https://doi.org/10.1056/NEJMOA0903840.

Maridas, D. E., Rendina-Ruedy, E., Helderman, R. C., DeMambro, V. E., Brooks, D., Guntur, A. R., et al. (2019). Progenitor recruitment and adipogenic lipolysis contribute to the anabolic actions of parathyroid hormone on the skeleton. *The FASEB Journal, 33*(2), 2885–2898. https://doi.org/10.1096/FJ.201800948RR.

McDonnell, E., Crown, S. B., Fox, D. B., Kitir, B., Ilkayeva, O. R., Olsen, C. A., et al. (2016). Lipids reprogram metabolism to become a major carbon source for histone acetylation. *Cell Reports, 17*(6), 1463–1472. https://doi.org/10.1016/J.CELREP.2016.10.012.

Méndez-Ferrer, S., Bonnet, D., Steensma, D. P., Hasserjian, R. P., Ghobrial, I. M., Gribben, J. G., et al. (2020). Bone marrow niches in haematological malignancies. *Nature Reviews Cancer*. https://doi.org/10.1038/s41568-020-0245-2.

Méndez-Ferrer, S., Lucas, D., Battista, M., & Frenette, P. S. (2008). Haematopoietic stem cell release is regulated by circadian oscillations. *Nature, 452*(7186), 442–447. https://doi.org/10.1038/NATURE06685.

Méndez-Ferrer, S., Michurina, T. V., Ferraro, F., Mazloom, A. R., MacArthur, B. D., Lira, S. A., et al. (2010). Mesenchymal and haematopoietic stem cells form a unique bone marrow niche. *Nature, 466*(7308), 829–834. https://doi.org/10.1038/nature09262.

Mesbahi, Y., Trahair, T. N., Lock, R. B., & Connerty, P. (2022). Exploring the metabolic landscape of AML: From Haematopoietic stem cells to Myeloblasts and Leukaemic stem cells. *Frontiers in Oncology, 12*, 281. https://doi.org/10.3389/FONC.2022.807266/BIBTEX.

Mistry, J. J., Marlein, C. R., Moore, J. A., Hellmich, C., Wojtowicz, E. E., Smith, J. G. W., et al. (2019). ROS-mediated PI3K activation drives mitochondrial transfer from stromal cells to hematopoietic stem cells in response to infection. *Proceedings of the National Academy of Sciences of the United States of America, 116*(49), 24610–24619. https://doi.org/10.1073/PNAS.1913278116.

Mitroulis, I., Chen, L. S., Singh, R. P., Kourtzelis, I., Economopoulou, M., Kajikawa, T., et al. (2017). Secreted protein Del-1 regulates myelopoiesis in the hematopoietic stem cell niche. *The Journal of Clinical Investigation, 127*(10), 3624. https://doi.org/10.1172/JCI92571.

Miyamoto, K., Araki, K. Y., Naka, K., Arai, F., Takubo, K., Yamazaki, S., et al. (2007). Foxo3a is essential for maintenance of the hematopoietic stem cell pool. *Cell Stem Cell, 1*(1), 101–112. https://doi.org/10.1016/J.STEM.2007.02.001.

Mizoguchi, T., Pinho, S., Ahmed, J., Kunisaki, Y., Hanoun, M., Mendelson, A., et al. (2014). Osterix Marks distinct waves of primitive and definitive stromal progenitors during bone marrow development. *Developmental Cell, 29*(3), 340–349. https://doi.org/10.1016/J.DEVCEL.2014.03.013.

Mojca, J. Z., Ro Zman, P. Z., Ivanovi, C. Z., & Bas, T. (1999). Concise review: The role of oxygen in hematopoietic stem cell physiology. *Journal of Cellular Physiology, 230*. https://doi.org/10.1002/jcp.24953.

Montecino-Rodriguez, E., Kong, Y., Casero, D., Rouault, A., Dorshkind, K., & Pioli, P. D. (2019). Lymphoid-biased hematopoietic stem cells are maintained with age and efficiently generate lymphoid progeny. *Stem Cell Reports, 12*(3), 584–596. https://doi.org/10.1016/J.STEMCR.2019.01.016.

Morrison, S. J., & Scadden, D. T. (2014). The bone marrow niche for haematopoietic stem cells. *Nature, 505*(7483), 327–334. NIH Public Access. https://doi.org/10.1038/nature12984.

Moschoi, R., Imbert, V., Nebout, M., Chiche, J., Mary, D., Prebet, T., et al. (2016). Protective mitochondrial transfer from bone marrow stromal cells to acute myeloid leukemic cells during chemotherapy. *Blood, 128*(2), 253–264. https://doi.org/10.1182/blood-2015-07-655860.

Nagai, M., Noguchi, R., Takahashi, D., Morikawa, T., Koshida, K., Komiyama, S., et al. (2019). Fasting-refeeding impacts immune cell dynamics and mucosal immune responses. *Cell, 178*(5), 1072–1087.e14. https://doi.org/10.1016/J.CELL.2019.07.047.

Nagareddy, P. R., Kraakman, M., Masters, S. L., Stirzaker, R. A., Gorman, D. J., Grant, R. W., et al. (2014). Adipose tissue macrophages promote Myelopoiesis and Monocytosis in obesity. *Cell Metabolism, 19*(5), 821–835. https://doi.org/10.1016/J.CMET.2014.03.029.

Nagareddy, P. R., Murphy, A. J., Stirzaker, R. A., Hu, Y., Yu, S., Miller, R. G., et al. (2013). Hyperglycemia promotes myelopoiesis and impairs the resolution of atherosclerosis. *Cell Metabolism, 17*(5), 695. https://doi.org/10.1016/J.CMET.2013.04.001.

Nakamura-Ishizu, A., Ito, K., & Suda, T. (2020). Hematopoietic stem cell metabolism during development and aging. *Developmental Cell, 54*(2), 239–255. https://doi.org/10.1016/j.devcel.2020.06.029.

Nakamura-Ishizu, A., Takubo, K., Fujioka, M., & Suda, T. (2014). Megakaryocytes are essential for HSC quiescence through the production of thrombopoietin. *Biochemical and Biophysical Research Communications, 454*(2), 353–357. https://doi.org/10.1016/J.BBRC.2014.10.095.

Naveiras, O., Nardi, V., Wenzel, P. L., Hauschka, P. V., Fahey, F., & Daley, G. Q. (2009). LETTERS bone-marrow adipocytes as negative regulators of the haematopoietic microenvironment. *Nature, 460*. https://doi.org/10.1038/nature08099.

Nelson, M. A. M., McLaughlin, K. L., Hagen, J. T., Coalson, H. S., Schmidt, C., Kassai, M., et al. (2021). Intrinsic OXPHOS limitations underlie cellular bioenergetics in leukemia. *eLife, 10*. https://doi.org/10.7554/ELIFE.63104.

Nguyen, T. M., Arthur, A., & Gronthos, S. (2016). The role of Eph/ephrin molecules in stromal–hematopoietic interactions. *International Journal of Hematology, 103*(2), 145–154. https://doi.org/10.1007/S12185-015-1886-X/FIGURES/4.

Nilsson, S. K., Johnston, H. M., Whitty, G. A., Williams, B., Webb, R. J., Denhardt, D. T., et al. (2005). Osteopontin, a key component of the hematopoietic stem cell niche and regulator of primitive hematopoietic progenitor cells. *Blood, 106*(4), 1232–1239. https://doi.org/10.1182/BLOOD-2004-11-4422.

Ning, K., Liu, S., Yang, B., Wang, R., Man, G., Wang, D., et al. (2022). Update on the effects of energy metabolism in bone marrow mesenchymal stem cells differentiation. *Molecular Metabolism, 58*. https://doi.org/10.1016/J.MOLMET.2022.101450.

Nobre, A. R., Risson, E., Singh, D. K., Di Martino, J. S., Cheung, J. F., Wang, J., et al. (2021). Bone marrow NG2 + /nestin + mesenchymal stem cells drive DTC dormancy via TGF-β2. *Nat Cancer*. https://doi.org/10.1038/s43018-021-00179-8.

Oguro, H., Ding, L., & Morrison, S. J. (2013). SLAM family markers resolve functionally distinct subpopulations of hematopoietic stem cells and multipotent progenitors. *Cell Stem Cell, 13*(1), 102–116. https://doi.org/10.1016/J.STEM.2013.05.014.

Orsini, M., Chateauvieux, S., Rhim, J., Gaigneaux, A., Cheillan, D., Christov, C., et al. (2019). Sphingolipid-mediated inflammatory signaling leading to autophagy inhibition converts erythropoiesis to myelopoiesis in human hematopoietic stem/progenitor cells. *Cell Death and Differentiation, 26*(9), 1796. https://doi.org/10.1038/S41418-018-0245-X.

Panaroni, C., & Wu, J. Y. (2013). Interactions between B lymphocytes and the osteoblast lineage in bone marrow. *Calcified Tissue International, 93*(3), 261–268. https://doi.org/10.1007/s00223-013-9753-3.

Park, I. K., Qian, D., Kiel, M., Becker, M. W., Pihalja, M., Weissman, I. L., et al. (2003). Bmi-1 is required for maintenance of adult self-renewing haematopoietic stem cells. *Nature, 423*(6937), 302–305. https://doi.org/10.1038/nature01587.

Passaro, D., Garcia-Albornoz, M., Diana, G., Malanchi, I., Gribben, J., Correspondence, D. B., et al. (2021). Integrated OMICs unveil the bone-marrow microenvironment in human leukemia. *Cell Reports, 35*. https://doi.org/10.1016/j.celrep.2021.109119.

Pernes, G., Flynn, M. C., Lancaster, G. I., & Murphy, A. J. (2019). Fat for fuel: Lipid metabolism in haematopoiesis. *Clinical & Translational Immunology, 8*(12), e1098. https://doi.org/10.1002/CTI2.1098.

Pietras, E. M., Mirantes-Barbeito, C., Fong, S., Loeffler, D., Kovtonyuk, L. V., Zhang, S., et al. (2016). Chronic interleukin-1 drives haematopoietic stem cells towards precocious myeloid differentiation at the expense of self-renewal. *Nature Cell Biology, 18*(6), 607. https://doi.org/10.1038/NCB3346.

Pinho, S., Marchand, T., Yang, E., Wei, Q., Nerlov, C., & Frenette, P. S. (2018). Lineage-biased hematopoietic stem cells are regulated by distinct niches. *Developmental Cell, 44*(5), 634–641.e4. https://doi.org/10.1016/j.devcel.2018.01.016.

Pittenger, M. F., Mackay, A. M., Beck, S. C., Jaiswal, R. K., Douglas, R., Mosca, J. D., et al. (1999). Multilineage potential of adult human mesenchymal stem cells. *Science, 284*(5411), 143–147. https://doi.org/10.1126/science.284.5411.143/suppl_file/983855s5_thumb.gif.

Pronk, E., & Raaijmakers, M. H. G. P. (2019). The mesenchymal niche in MDS. *Blood, 133*(10), 1031–1038. American Society of Hematology. https://doi.org/10.1182/blood-2018-10-844639.

Puneet Agarwal, A., Isringhausen, S., & Li, H. (2019). Mesenchymal niche-specific expression of Cxcl12 controls quiescence of treatment-resistant leukemia stem cells. *Cell Stem Cell*. https://doi.org/10.1016/j.stem.2019.02.018.

Raaijmakers, M. H. G. P. (2012). Myelodysplastic syndromes: Revisiting the role of the bone marrow microenvironment in disease pathogenesis. *International Journal of Hematology, 95*(1), 17–25. https://doi.org/10.1007/s12185-011-1001-x.

Raaijmakers, M. H. G. P., Mukherjee, S., Guo, S., Zhang, S., Kobayashi, T., Schoonmaker, J. A., et al. (2010). Bone progenitor dysfunction induces myelodysplasia and secondary leukaemia. *Nature, 464*(7290), 852–857. https://doi.org/10.1038/nature08851.

Rashidi, N. M., Scott, M. K., Scherf, N., Krinner, A., Kalchschmidt, J. S., Gounaris, K., et al. (2014). In vivo time-lapse imaging shows diverse niche engagement by quiescent and naturally activated hematopoietic stem cells. *Blood, 124*(1), 79–83. https://doi.org/10.1182/blood-2013-10-534859.

Ren, R., Ocampo, A., Liu, G.-H., Carlos, J., & Belmonte, I. (2017). Regulation of stem cell aging by metabolism and epigenetics. *Cell Metabolism*. https://doi.org/10.1016/j.cmet.2017.07.019.

Rendina-Ruedy, E., & Rosen, C. J. (2020). Lipids in the bone marrow: An evolving perspective. *Cell Metabolism, 31*(2), 219–231. https://doi.org/10.1016/J.CMET.2019.09.015.

Robbins, C. S., Chudnovskiy, A., Rauch, P. J., Figueiredo, J. L., Iwamoto, Y., Gorbatov, R., et al. (2012). Extramedullary hematopoiesis generates Ly-6C high monocytes that infiltrate atherosclerotic lesions. *Circulation, 125*(2), 364–374. https://doi.org/10.1161/CIRCULATIONAHA.111.061986/FORMAT/EPUB.

Ruggero, D., & Pandolfi, P. P. (2003). Does the ribosome translate cancer? *Nature Reviews Cancer, 3*(3), 179–192. https://doi.org/10.1038/nrc1015.

Rupec, R. A., Jundt, F., Rebholz, B., Eckelt, B., Weindl, G., Herzinger, T., et al. (2005). Stroma-mediated dysregulation of Myelopoiesis in mice lacking IκBα. *Immunity*, *22*(4), 479–491. https://doi.org/10.1016/J.IMMUNI.2005.02.009.

Saçma, M., Pospiech, J., Bogeska, R., de Back, W., Mallm, J. P., Sakk, V., et al. (2019). Haematopoietic stem cells in perisinusoidal niches are protected from ageing. *Nature Cell Biology*, *21*(11), 1309–1320. https://doi.org/10.1038/s41556-019-0418-y.

Saito, Y., Chapple, R. H., Lin, A., Kitano, A., & Nakada, D. (2015). AMPK protects leukemia-initiating cells in myeloid leukemias from metabolic stress in the bone marrow. *Cell Stem Cell*, *17*(5), 585. https://doi.org/10.1016/J.STEM.2015.08.019.

Saito, K., Zhang, Q., Yang, H., Yamatani, K., Ai, T., Ruvolo, V., et al. (2021). Exogenous mitochondrial transfer and endogenous mitochondrial fission facilitate AML resistance to OxPhos inhibition. *Blood Advances*, *5*(20), 4233–4255. https://doi.org/10.1182/BLOODADVANCES.2020003661.

Sánchez-Aguilera, A., & Méndez-Ferrer, S. (2017). The hematopoietic stem-cell niche in health and leukemia. *Cellular and Molecular Life Sciences: CMLS*, *74*(4), 579–590. https://doi.org/10.1007/s00018-016-2306-y.

Santos, A. X. S., Maia, J. E., Crespo, P. M., Pettenuzzo, L. F., Daniotti, J. L., Barbé-Tuana, F. M., et al. (2011). GD1a modulates GM-CSF-induced cell proliferation. *Cytokine*, *56*(3), 600–607. https://doi.org/10.1016/J.CYTO.2011.08.032.

Schmidt, E. V. (1999). The role of c-myc in cellular growth control. *Oncogene*, *18*(19), 2988–2996. https://doi.org/10.1038/sj.onc.1202751.

Schofield, R. (1978). The relationship between the spleen colony-forming cell and the haemopoietic stem cell. *Blood Cells*, *4*, 7–25.

Schüler, S. C., Gebert, N., & Ori, A. (2020). Stem cell aging: The upcoming era of proteins and metabolites. *Mechanisms of Ageing and Development*, *190*, 111288. https://doi.org/10.1016/J.MAD.2020.111288.

Seijkens, T., Hoeksema, M. A., Beckers, L., Smeets, E., Meiler, S., Levels, J., et al. (2014). Hypercholesterolemia-induced priming of hematopoietic stem and progenitor cells aggravates atherosclerosis. *The FASEB Journal*, *28*(5), 2202–2213. https://doi.org/10.1096/FJ.13-243105.

Shafat, M. S., Oellerich, T., Mohr, S., Robinson, S. D., Edwards, D. R., Marlein, C. R., et al. (2017). Leukemic blasts program bone marrow adipocytes to generate a protumoral microenvironment. *Blood*, *129*(10), 1320–1332. https://doi.org/10.1182/BLOOD-2016-08-734798.

Shen, B., Tasdogan, A., Ubellacker, J. M., Zhang, J., Nosyreva, E. D., Du, L., et al. (2021). A mechanosensitive peri-arteriolar niche for osteogenesis and lymphopoiesis. *Nature*, *591*(7850), 438–444. https://doi.org/10.1038/s41586-021-03298-5.

Shi, K., Li, H., Chang, T., He, W., Kong, Y., Qi, C., et al. (2022). Bone marrow hematopoiesis drives multiple sclerosis progression. *Cell*. https://doi.org/10.1016/j.cell.2022.05.020.

Signer, R. A. J., Magee, J. A., Salic, A., & Morrison, S. J. (2014). Haematopoietic stem cells require a highly regulated protein synthesis rate. *Nature*, *508*(7498), 49–54. https://doi.org/10.1038/nature13035.

Signer, R. A. J., Qi, L., Zhao, Z., Thompson, D., Sigova, A. A., Fan, Z. P., et al. (2016). The rate of protein synthesis in hematopoietic stem cells is limited partly by 4E-BPs. *Genes and Development*, *30*(15), 1698–1703. https://doi.org/10.1101/GAD.282756.116/-/DC1.

Siminovitch, L., Mcculloch, E. A., & Till, J. E. (1963). The distribution of colony-forming cells among spleen colonies. *Journal of Cellular and Comparative Physiology*, *62*, 327–336. https://doi.org/10.1002/JCP.1030620313.

Siminovitch, L., Till, J. E., & McCulloch, E. A. (1964). Decline in colony-forming ability of marrow cells subjected to serial transplantation into irradiated mice. *Journal of Cellular and Comparative Physiology*, *64*(1), 23–31. https://doi.org/10.1002/JCP.1030640104.

Simons, K., & Ehehalt, R. (2002). Cholesterol, lipid rafts, and disease. *The Journal of Clinical Investigation, 110*(5), 597–603. https://doi.org/10.1172/JCI16390.

Simsek, T., Kocabas, F., Zheng, J., Deberardinis, R. J., Mahmoud, A. I., Olson, E. N., et al. (2010). The distinct metabolic profile of hematopoietic stem cells reflects their location in a hypoxic niche. *Cell Stem Cell,* 7(3), 380–390. https://doi.org/10.1016/J.STEM.2010.07.011.

Singer, K., DelProposto, J., Lee Morris, D., Zamarron, B., Mergian, T., Maley, N., et al. (2014). Diet-induced obesity promotes myelopoiesis in hematopoietic stem cells. *Molecular Metabolism, 3*(6), 664. https://doi.org/10.1016/J.MOLMET.2014.06.005.

Singh, A. K., & Cancelas, J. A. (2021). Mitochondria transfer in bone marrow hematopoietic activity. *Current Stem Cell Reports,* 7(1), 1. https://doi.org/10.1007/S40778-020-00185-Z.

Škrtić, M., Sriskanthadevan, S., Jhas, B., Gebbia, M., Wang, X., Wang, Z., et al. (2011). Inhibition of mitochondrial translation as a therapeutic strategy for human acute myeloid leukemia. *Cancer Cell, 20*(5), 674–688. https://doi.org/10.1016/J.CCR.2011.10.015.

Solimando, A. G., Melaccio, A., Vacca, A., & Ria, R. (2022). The bone marrow niche landscape: A journey through aging, extrinsic and intrinsic stressors in the haemopoietic milieu. *Journal of Cancer Metastasis and Treatment, 8*, 9. https://doi.org/10.20517/2394-4722.2021.166.

Song, K., Li, M., Xu, X., Xuan, L., Huang, G., & Liu, Q. (2016). Resistance to chemotherapy is associated with altered glucose metabolism in acute myeloid leukemia. *Oncology Letters, 12*(1), 334–342. https://doi.org/10.3892/OL.2016.4600/DOWNLOAD.

Spevak, C. C., Elias, H. K., Kannan, L., Ali, M. A. E., Martin, G. H., Selvaraj, S., et al. (2020). Hematopoietic stem and progenitor cells exhibit stage-specific translational programs via mTOR- and CDK1-dependent mechanisms. *Cell Stem Cell, 26*(5), 755–765.e7. https://doi.org/10.1016/j.stem.2019.12.006.

Sriskanthadevan, S., Jeyaraju, D. V., Chung, T. E., Prabha, S., Xu, W., Skrtic, M., et al. (2015). AML cells have low spare reserve capacity in their respiratory chain that renders them susceptible to oxidative metabolic stress. *Blood, 125*(13), 2120. https://doi.org/10.1182/BLOOD-2014-08-594408.

Stevens, B. M., Jones, C. L., Pollyea, D. A., Culp-Hill, R., D'Alessandro, A., Winters, A., et al. (2020). Fatty acid metabolism underlies venetoclax resistance in acute myeloid leukemia stem cells. *Nature Cancer, 1*(12), 1176–1187. https://doi.org/10.1038/s43018-020-00126-z.

Sulima, S. O., & Keersmaecker, K. D. (2018). Bloody mysteries of ribosomes. *Hemasphere*. https://doi.org/10.1097/HS9.0000000000000095.

Swirski, F. K., Libby, P., Aikawa, E., Alcaide, P., Luscinskas, F. W., Weissleder, R., et al. (2007). Ly-6Chi monocytes dominate hypercholesterolemia-associated monocytosis and give rise to macrophages in atheromata. *The Journal of Clinical Investigation, 117*(1), 195–205. https://doi.org/10.1172/JCI29950.

Tabe, Y., Yamamoto, S., Saitoh, K., Sekihara, K., Monma, N., Ikeo, K., et al. (2017). Bone marrow adipocytes facilitate fatty acid oxidation activating AMPK and a transcriptional network supporting survival of acute monocytic leukemia cells. *Cancer Research,* 77(6), 1453–1464. https://doi.org/10.1158/0008-5472.CAN-16-1645.

Tadokoro, Y., & Hirao, A. (2022). The role of nutrients in maintaining hematopoietic stem cells and healthy hematopoiesis for life. *International Journal Of Molecular Sciences, 23*(3), *1574*. https://doi.org/10.3390/IJMS23031574.

Takubo, K., Goda, N., Yamada, W., Iriuchishima, H., Ikeda, E., Kubota, Y., et al. (2010). Regulation of the HIF-1α level is essential for hematopoietic stem cells. *Cell Stem Cell,* 7(3), 391–402. https://doi.org/10.1016/J.STEM.2010.06.020.

Takubo, K., Nagamatsu, G., Kobayashi, C. I., Nakamura-Ishizu, A., Kobayashi, H., Ikeda, E., et al. (2013). Regulation of glycolysis by Pdk functions as a metabolic checkpoint for cell cycle quiescence in hematopoietic stem cells. *Cell Stem Cell, 12*(1), 49–61. https://doi.org/10.1016/J.STEM.2012.10.011.

Tamplin, O. J., Durand, E. M., Carr, L. A., Childs, S. J., Hagedorn, E. J., Li, P., et al. (2015). Hematopoietic stem cell arrival triggers dynamic remodeling of the perivascular niche. *Cell, 160*(1–2), 241–252. https://doi.org/10.1016/J.CELL.2014.12.032.

Tang, D., Tao, S., Chen, Z., Koliesnik, I. O., Calmes, P. G., Hoerr, V., et al. (2016). Dietary restriction improves repopulation but impairs lymphoid differentiation capacity of hematopoietic stem cells in early aging. *Journal of Experimental Medicine, 213*(4), 535–553. https://doi.org/10.1084/JEM.20151100.

Taya, Y., Ota, Y., Wilkinson, A. C., Kanazawa, A., Watarai, H., Kasai, M., et al. (2016). Depleting dietary valine permits nonmyeloablative mouse hematopoietic stem cell transplantation. *Science (New York, N.Y.), 354*(6316), 1152–1155. https://doi.org/10.1126/SCIENCE.AAG3145.

Tcheng, M., Roma, A., Ahmed, N., Smith, R. W., Jayanth, P., Minden, M. D., et al. (2021). Very long chain fatty acid metabolism is required in acute myeloid leukemia. *Blood, 137*(25), 3518–3532. https://doi.org/10.1182/BLOOD.2020008551.

Tie, G., Messina, K. E., Yan, J., Messina, J. A., & Messina, L. M. (2014). Hypercholesterolemia induces oxidant stress that accelerates the ageing of hematopoietic stem cells. *Journal of the American Heart Association, 3*(1). https://doi.org/10.1161/JAHA.113.000241.

Tikhonova, A. N., Dolgalev, I., Hu, H., Sivaraj, K. K., Hoxha, E., Cuesta-Domínguez, Á., et al. (2019). The bone marrow microenvironment at single-cell resolution. *Nature, 569*(7755), 222–228. https://doi.org/10.1038/s41586-019-1104-8.

Tirado, H. A., Balasundaram, N., Laaouimir, L., Erdem, A., & van Gastel, N. (2023). Metabolic crosstalk between stromal and malignant cells in the bone marrow niche. *Bone Reports, 18*, 101669. https://doi.org/10.1016/J.BONR.2023.101669.

Tong, H., Hu, C., Zhuang, Z., Wang, L., & Jin, J. (2012). Hypoxia-inducible factor-1α expression indicates poor prognosis in myelodysplastic syndromes. *Leukemia and Lymphoma, 53*(12), 2412–2418. https://doi.org/10.3109/10428194.2012.696637/SUPPL_FILE/DISCLOSURE.ZIP.

Tothova, Z., Kollipara, R., Huntly, B. J., Lee, B. H., Castrillon, D. H., Cullen, D. E., et al. (2007). FoxOs are critical mediators of hematopoietic stem cell resistance to physiologic oxidative stress. *Cell, 128*(2), 325–339. https://doi.org/10.1016/J.CELL.2007.01.003.

Tratwal, J., Rojas-Sutterlin, S., Bataclan, C., Blum, S., & Naveiras, O. (2021). Bone marrow adiposity and the hematopoietic niche: A historical perspective of reciprocity, heterogeneity, and lineage commitment. *Best Practice & Research. Clinical Endocrinology & Metabolism, 35*(4). https://doi.org/10.1016/J.BEEM.2021.101564.

Tsai, C.-H., Chuang, Y.-M., Lunt, S. Y., Kaech, S. M., & Ho Correspondence, P.-C. (2023). Immunoediting instructs tumor metabolic reprogramming to support immune evasion. *Cell Metab*. https://doi.org/10.1016/j.cmet.2022.12.003.

Umemoto, T., Johansson, A., Adil, S., Ahmad, I., Hashimoto, M., Kubota, S., et al. (2022). ATP citrate lyase controls hematopoietic stem cell fate and supports bone marrow regeneration. *The EMBO Journal, 41*(8), e109463. https://doi.org/10.15252/EMBJ.2021109463.

Unwin, R. D., Smith, D. L., Blinco, D., Wilson, C. L., Miller, C. J., Evans, C. A., et al. (2006). Quantitative proteomics reveals posttranslational control as a regulatory factor in primary hematopoietic stem cells. *Blood, 107*(12), 4687–4694. https://doi.org/10.1182/BLOOD-2005-12-4995.

Valencia, T., Kim, J. Y., Abu-Baker, S., Moscat-Pardos, J., Ahn, C. S., Reina-Campos, M., et al. (2014). Metabolic reprogramming of stromal fibroblasts through p62-mTORC1 signaling promotes inflammation and tumorigenesis. *Cancer Cell, 26*(1), 121–135. https://doi.org/10.1016/j.ccr.2014.05.004.

Van Den Berg, S. M., Seijkens, T. T., HKusters, P. J., Beckers, L., DenToom, M., Smeets, E., et al. (2016). Diet-induced obesity in mice diminishes hematopoietic stem and progenitor cells in the bone marrow. *The FASEB Journal*, *30*(5), 1779–1788. https://doi.org/10.1096/FJ.201500175.

van Galen, P., Mbong, N., Kreso, A., Schoof, E. M., Wagenblast, E., Ng, S. W. K., et al. (2018). Integrated stress response activity Marks stem cells in Normal hematopoiesis and leukemia. *Cell Reports*, *25*(5), 1109–1117.e5. https://doi.org/10.1016/j.celrep.2018.10.021.

van Gastel, N., Spinelli, J. B., Sharda, A., Schajnovitz, A., Baryawno, N., Rhee, C., et al. (2020). Induction of a timed metabolic collapse to overcome Cancer Chemoresistance. *Cell Metabolism*. https://doi.org/10.1016/j.cmet.2020.07.009.

Vilaplana-Lopera, N., Cuminetti, V., Almaghrabi, R., Papatzikas, G., Rout, A. K., Jeeves, M., et al. (2022). Crosstalk between AML and stromal cells triggers acetate secretion through the metabolic rewiring of stromal cells. *eLife*, *11*. https://doi.org/10.7554/ELIFE.75908.

Viñado, A. C., Calvo, I. A., Cenzano, I., Olaverri, D., Cocera, M., San Martin-Uriz, P., et al. (2022). The bone marrow niche regulates redox and energy balance in MLL::AF9 leukemia stem cells. *Leukemia 2022 36:8*, *36*(8), 1969–1979. https://doi.org/10.1038/s41375-022-01601-5.

Vukovic, M., Guitart, A. V., Sepulveda, C., Villacreces, A., O'Duibhir, E., Panagopoulou, T. I., et al. (2015). Hif-1α and Hif-2α synergize to suppress AML development but are dispensable for disease maintenance. *The Journal of Experimental Medicine*, *212*(13), 2223–2234. https://doi.org/10.1084/JEM.20150452.

Waite, K. J., Floyd, Z. E., Arbour-Reily, P., & Stephens, J. M. (2000). Interferon-induced regulation of peroxisome proliferator-activated receptor and STATs in adipocytes. *J Biol Chem*. https://doi.org/10.1074/jbc.M007894200.

Walkley, C. R., Olsen, G. H., Dworkin, S., Fabb, S. A., Swann, J., McArthur, G. A. A., et al. (2007). A microenvironment-induced myeloproliferative syndrome caused by retinoic acid receptor gamma deficiency. *Cell*, *129*(6), 1097–1110. https://doi.org/10.1016/J.CELL.2007.05.014.

Walkley, C. R., Shea, J. M., Sims, N. A., Purton, L. E., & Orkin, S. H. (2007). Rb regulates interactions between hematopoietic stem cells and their bone marrow microenvironment. *Cell*, *129*(6), 1081–1095. https://doi.org/10.1016/J.CELL.2007.03.055.

Wang, L., Benedito, R., Bixel, M. G., Zeuschner, D., Stehling, M., Sävendahl, L., et al. (2013). Identification of a clonally expanding haematopoietic compartment in bone marrow. *The EMBO Journal*, *32*(2), 219. https://doi.org/10.1038/EMBOJ.2012.308.

Wang, X., Cooper, S., Broxmeyer, H. E., & Kapur, R. (2022). Nuclear translocation of TFE3 under hypoxia enhances the engraftment of human hematopoietic stem cells. *Leukemia*. https://doi.org/10.1038/s41375-022-01628-8.

Wang, Y.-H., Israelsen, W. J., Lee, D., Yu, V. W. C., Jeanson, N. T., Clish, C. B., et al. (2014). Cell-state-specific metabolic dependency in hematopoiesis and Leukemogenesis. *Cell*. https://doi.org/10.1016/j.cell.2014.07.048.

Wei, Q., & Frenette, P. S. (2018). Niches for hematopoietic stem cells and their progeny. *Immunity*, *48*(4), 632–648. https://doi.org/10.1016/J.IMMUNI.2018.03.024.

Wierenga, A. T. J., Cunningham, A., Erdem, A., Lopera, N. V., Brouwers-Vos, A. Z., Pruis, M., et al. (2019). HIF1/2-exerted control over glycolytic gene expression is not functionally relevant for glycolysis in human leukemic stem/progenitor cells. *Cancer & Metabolism*, 7(1), 1–17. https://doi.org/10.1186/S40170-019-0206-Y.

Wilkinson, A. C., Morita, M., Nakauchia, H., & Yamazaki, S. (2018). Branched-chain amino acid depletion conditions bone marrow for hematopoietic stem cell transplantation avoiding amino acid imbalance-associated toxicity. *Experimental Hematology*, *63*, 12–16.e1. https://doi.org/10.1016/J.EXPHEM.2018.04.004.

Willems, L., Jacque, N., Jacquel, A., Neveux, N., Maciel, T. T., Lambert, M., et al. (2013). Inhibiting glutamine uptake represents an attractive new strategy for treating acute myeloid leukemia. *Blood*, *122*(20), 3521. https://doi.org/10.1182/BLOOD-2013-03-493163.

Wilson, A., Fu, H., Schiffrin, M., Winkler, C., Koufany, M., Jouzeau, J. Y., et al. (2018). Lack of adipocytes alters hematopoiesis in Lipodystrophic mice. *Frontiers in Immunology*, *9*. https://doi.org/10.3389/FIMMU.2018.02573.

Wilson, A., Laurenti, E., Oser, G., Van Der Wath, R. C., Blanco-Bose, W., Jaworski, M., et al. (2008). Hematopoietic stem cells reversibly switch from dormancy to self-renewal during homeostasis and repair. *Cell*. https://doi.org/10.1016/j.cell.2008.10.048.

Xie, Y., Yin, T., Wiegraebe, W., He, X. C., Miller, D., Stark, D., et al. (2009). Detection of functional haematopoietic stem cell niche using real-time imaging. *Nature*, *457*(7225), 97–101. https://doi.org/10.1038/NATURE07639.

Xu, B., Hu, R., Liang, Z., Chen, T., Chen, J., Hu, Y., et al. (2021). Metabolic regulation of the bone marrow microenvironment in leukemia. *Blood Reviews*, *48*, 100786. https://doi.org/10.1016/J.BLRE.2020.100786.

Xu, Y., Murphy, A. J., & Fleetwood, A. J. (2022). Hematopoietic progenitors and the bone marrow niche shape the inflammatory response and contribute to chronic disease. *International Journal of Molecular Sciences*, *23*(4). https://doi.org/10.3390/IJMS23042234.

Yamazaki, S., Iwama, A., Takayanagi, S. I., Morita, Y., Eto, K., Ema, H., et al. (2006). Cytokine signals modulated via lipid rafts mimic niche signals and induce hibernation in hematopoietic stem cells. *The EMBO Journal*, *25*(15), 3515. https://doi.org/10.1038/SJ.EMBOJ.7601236.

Yao, L., Yokota, T., Xia, L., Kincade, P. W., & McEver, R. P. (2005). Bone marrow dysfunction in mice lacking the cytokine receptor gp130 in endothelial cells. *Blood*, *106*(13), 4093–4101. https://doi.org/10.1182/BLOOD-2005-02-0671.

Ye, H., Adane, B., Stranahan, A. W., Park, C. Y., & Correspondence, C. T. J. (2016). Leukemic stem cells evade chemotherapy by metabolic adaptation to an adipose tissue niche. *Cell Stem Cell*, *19*, 23–37. https://doi.org/10.1016/j.stem.2016.06.001.

Yeh, S. C. A., Hou, J., Wu, J. W., Yu, S., Zhang, Y., Belfield, K. D., et al. (2022). Quantification of bone marrow interstitial pH and calcium concentration by intravital ratiometric imaging. *Nature Communications*, *13*(1). https://doi.org/10.1038/S41467-022-27973-X.

Yu, W. M., Liu, X., Shen, J., Jovanovic, O., Pohl, E. E., Gerson, S. L., et al. (2013). Metabolic regulation by the mitochondrial phosphatase PTPMT1 is required for hematopoietic stem cell differentiation. *Cell Stem Cell*, *12*(1), 62–74. https://doi.org/10.1016/J.STEM.2012.11.022.

Yue, R., Zhou, B. O., Shimada, I. S., Zhao, Z., & Morrison, S. J. (2016). Leptin receptor promotes Adipogenesis and reduces osteogenesis by regulating mesenchymal stromal cells in adult bone marrow. *Cell Stem Cell*, *18*(6), 782–796. https://doi.org/10.1016/J.STEM.2016.02.015.

Yuzefovych, L., Wilson, G., & Rachek, L. (2010). Different effects of oleate vs. palmitate on mitochondrial function, apoptosis, and insulin signaling in L6 skeletal muscle cells: Role of oxidative stress. *American Journal of Physiology - Endocrinology and Metabolism*, *299*(6), 1096–1105. https://doi.org/10.1152/AJPENDO.00238.2010/ASSET/IMAGES/LARGE/ZH10121061190009.JPEG.

Zahr, A. A., Salama, M. E., Carreau, N., Tremblay, D., Verstovsek, S., Mesa, R., et al. (2016). Bone marrow fibrosis in myelofibrosis: Pathogenesis, prognosis and targeted strategies. *Haematologica*, *101*(6), 660. https://doi.org/10.3324/HAEMATOL.2015.141283.

Zhang, Z., Huang, Z., Ong, B., Sahu, C., Zeng, H., & Ruan, H. B. (2019). Bone marrow adipose tissue-derived stem cell factor mediates metabolic regulation of hematopoiesis. *Haematologica*, *104*(9), 1731–1743. https://doi.org/10.3324/haematol.2018.205856.
Zhang, H., Liesveld, J. L., Calvi, L. M., Lipe, B. C., Xing, L., Becker, M. W., et al. (2023). The roles of bone remodeling in normal hematopoiesis and age-related hematological malignancies. *Bone Res*. https://doi.org/10.1038/s41413-023-00249-w.
Zhang, J., Niu, C., Ye, L., Huang, H., He, X., Tong, W. G., et al. (2003). Identification of the haematopoietic stem cell niche and control of the niche size. *Nature*, *425*(6960), 836–841. https://doi.org/10.1038/NATURE02041.
Zhang, C. C., & Sadek, H. A. (2014). Hypoxia and metabolic properties of hematopoietic stem cells. *Antioxidants and redox signaling*, *20*(12), 1891–1901. Mary Ann Liebert Inc https://doi.org/10.1089/ars.2012.5019.
Zhao, M., Perry, J. M., Marshall, H., Venkatraman, A., Qian, P., He, X. C., et al. (2014). Megakaryocytes maintain homeostatic quiescence and promote post-injury regeneration of hematopoietic stem cells. *Nature Medicine*, *20*(11), 1321–1326. https://doi.org/10.1038/nm.3706.
Zhou, K., Yao, P., He, J., & Zhao, H. (2019). Lipophagy in nonliver tissues and some related diseases: Pathogenic and therapeutic implications. *Journal of Cellular Physiology*, *234*(6), 7938–7947. https://doi.org/10.1002/JCP.27988.
Zhou, B. O., Yu, H., Yue, R., Zhao, Z., Rios, J. J., Naveiras, O., et al. (2017). Bone marrow adipocytes promote the regeneration of stem cells and haematopoiesis by secreting SCF. *Nature Cell Biology*, *19*(8), 891–903. https://doi.org/10.1038/NCB3570.
Ziulkoski, A. L., Andrade, C. M. B., Crespo, P. M., Sisti, E., Trindade, V. M. T., Daniotti, J. L., et al. (2006). Gangliosides of myelosupportive stroma cells are transferred to myeloid progenitors and are required for their survival and proliferation. *Biochemical Journal*, *394*(Pt 1), 1. https://doi.org/10.1042/BJ20051189.

Immune cell interactions with the stem cell niche

Etienne C.E. Wang*
National Skin Centre, Skin Research Institute of Singapore, Singapore, Singapore
*Corresponding author: e-mail address: etienne@nsc.com.sg

Contents

Advances in Stem Cells and their Niches, Volume 7
ISSN 2468-5097
https://doi.org/10.1016/bs.asn.2023.07.001

1. Introduction

The specific and conducive niche microenvironment is essential to the function and survival of stem cells *in vivo*. The niche was defined by Schofield in 1978 to describe the "association with other cells which determines [a stem cell's] behavior" (Schofield, 1978). The anatomical milieu of stem cells positions them to receive signals from paracrine factors, adhesion molecules and other cell–cell interactions that supports their longevity, quiescence, self-renewal and pluripotency (Fuchs, Tumbar, & Guasch, 2004; Goldstein & Horsley, 2012; Li & Xie, 2005). The niche also provides protection for stem cells from exogenous and endogenous insults such as pathogens, chemotherapy, and autoimmunity (Niederkorn & Stein-Streilein, 2010). Removing stem cells from their niche (e.g., by physiological competition, *ex vivo* or *in vitro* culture, or genetic methods to ablate portions of the niche) frequently leads to activation of cellular programs that push the cells toward proliferation and differentiation, and results in loss of stemness (Hsu, Pasolli, & Fuchs, 2011; Rompolas, Mesa, & Greco, 2013). The cells and structures that play a role in creating the stem cell niche are usually surrounding cells (epithelial or stromal), extracellular matrix (ECM), endothelial and neural cells, in a manner that is conserved across many organs like the bone marrow, gut and brain (Jones & Wagers, 2008). Comprehensive deconstruction and understand of stem cell–niche interactions is essential for the advancement of regeneration medicine, and our ability to produce viable tissue replacements *ex vivo* (Lane, Williams, & Watt, 2014). This chapter will review the role of immune cells in their interactions with the stem cell niche, the mechanisms through which they help to support stem cell activity, and how they might be harnessed in the treatment of disorders, particularly in the skin and hair.

2. Stem cell niches of the skin and hair follicle

The skin is the first line of defense from environmental insults, and an important organ involved in thermoregulation and preventing dehydration. Different epithelial compartments of the skin have different rates of turnover, and different responses to wounding, and this is reflected in the diverse stem cell compartments found in the skin and its appendages, which include hair, nails and sweat/sebaceous/apocrine glands (Fig. 1) (Braun & Prowse, 2006).

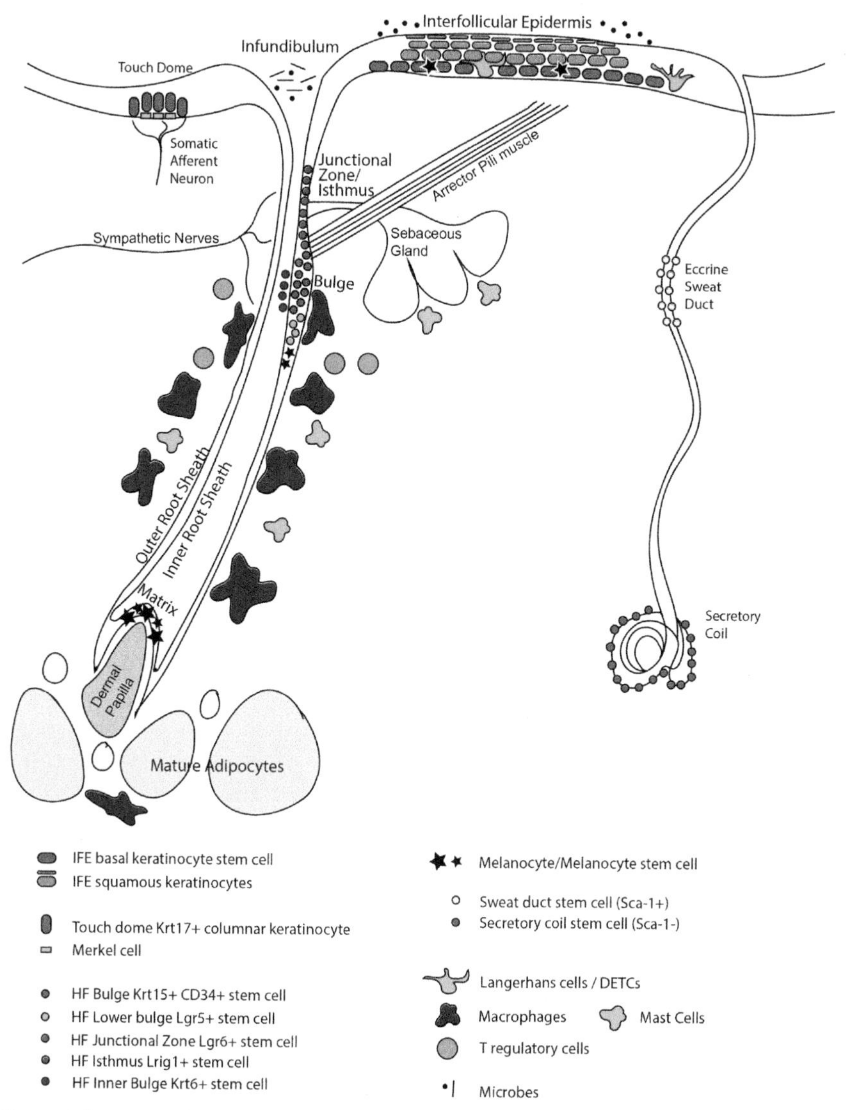

Fig. 1 Schematic of epidermal stem cells and their niche. Epidermal stem cells are found in the basal layer of the interfollicular epidermis (IFE), touch dome, sweat glands, and the hair follicles (HF). IFE stem cells give rise to keratinocytes that undergo a cornification program that produces the stratified epithelium, and the basal niche is comprised of the basement membrane, basal layer melanocytes/melanocyte stem cells, dendritic epidermal T cells (DETCs) and Langerhans cells. Specialized Krt17+ epidermal stem cells serve as progenitor cells for Merkel cells in touch domes, which confer mechanosensitivity to soft touch in the epidermis. Epidermal stem cells that are recruited into skin appendages take on more specialized roles. The eccrine sweat gland is formed from

(Continued)

2.1 Interfollicular epidermis

The interfollicular epidermis (IFE) is a stratified squamous epithelium that regenerates from the stochastic competitive stem cell pool in the basal layer (Blanpain & Simons, 2013). The progeny of this basal pool of Krt5+ Krt14+ keratinocyte stem cells are maintained by autocrine Wnt signals (Lim et al., 2013), and differentiate upwards and outwards when they leave the basal niche, and turns over approximately every 28 days (Halprin, 1972). Basement membrane ECM is a vital component of the IFE stem cell niche, and includes α6, β1, β4 integrins, and laminin-5 (Hsu, Li, & Fuchs, 2014). Collagen 17A in the basement membrane, which is associated with junctional epidermolysis bullosa and bullous pemphigoid, also regulates the proliferation of basal keratinocyte stem cells (Watanabe et al., 2017). Defects in the IFE are resolved by a wound-healing process that involves keratinocytes, stromal mesenchymal cells and immune cells. During the steady state, antigen-presenting cells like Langerhans cells are present within the epidermis where they function as immune sentinels (Merad, Ginhoux, & Collin, 2008).

2.2 Hair follicle

The hair follicle (HF) has several stem cell compartments that are tightly regulated in order to orchestrate the hair cycle, whereby each follicle undergoes phases of growth (anagen), regression (catagen) and rest (telogen) throughout the life of the mammal. The hierarchical nature of the stem cell niche in the HF reflects this function, with the HF Krt15+ bulge (CD34+ in the

Fig. 1—Cont'd direct invagination of the IFE during embryogenesis, with Sca-1+ stem cells in the duct and Sca-1− stem cells in the secretory coil. The ductal Sca-1+ cells are capable to contributing to wound healing in the IFE. Both ductal and glandular stem cells are further arranged radially as myoepithelial (outer) or luminal (inner) unipotent progenitors to give rise to cells that replenish their respective layers with almost no transdifferentiation. HF stem cells are also established during embryogenesis, with the most long-lived, slow-cycling stem cells in the bulge (CD34+). Other stem cells compartments in the HF include the inner bulge (Krt6+), lower bulge (Lgr5+), junctional zone/isthmus (Lrig1+/Lgr6+). The lower bulge cells give rise to outer root sheath keratinocytes that proliferate during anagen to give rise to the matrix of the HF bulb, where it undergoes epithelial–mesenchymal interactions with the dermal papilla (DP) to differentiate into specialized layers that form the growing hair shaft. The DP itself is replenished from stem cells likely arising from the dermal sheath (out of scope of this chapter). The HF stem cell niche is complex and has been shown to include mesenchymal cells like the DP, inner bulge, sympathetic nerve endings, smooth muscle attachments, adipocytes, and immune cells like macrophages, DETCs and T regs.

mouse) containing the most long-lived self-renewing population of stem cells. A precise balance of Wnt and BMP signaling in the niche is essential for the control of HF stem cell activity (Greco et al., 2009; Kobielak et al., 2007; Lien et al., 2014; Lim et al., 2016; Plikus et al., 2008). As stem cells leave the bulge, their stemness and pluripotency is reduced, and establish other stem cell populations in the lower bulge (Lgr5+), isthmus and infundibulum (Lrig1+), sebaceous duct (Lrig1 Lgr6+), and inner bulge (Krt6+) of the HF (Hsu et al., 2011; Kretzschmar & Watt, 2014). These stem cells are more readily mobilized to differentiate into keratinocytes for the outer root sheath and sebaceous ducts during physiological anagen (Hsu et al., 2011), and the IFE during wound healing (Ito et al., 2005). The HF stem cell niche in the bulge is made up of neighboring keratinocytes (i.e., the inner bulge), the basement membrane ECM, as well as other cell types in the local microenvironment such as the arrector pili muscle attachment and sympathetic nerves. Collagen 17A, which is important in IFE stem cell homeostasis, also plays a role in HF stem cell homeostasis (Matsumura et al., 2016). Stem cells in the lower bulge are juxtaposed to the dermal papilla (DP) during telogen, and give rise to keratinocytes of the outer root sheath during anagen, during which they extend into the adipose layer (Festa et al., 2011).

The HF stem cell compartment is a site of immune privilege, with a reduced expression of MHC molecules (Meyer et al., 2008). This is postulated to be required for the avoidance of autoimmune destruction of stem cell compartments (keratinocyte and melanocyte) at a site of increased microbial traffic (i.e., the HF infundibulum). This immune privilege also extends to the bulb region of the HF, and protects the HF matrix and DP as well. Immunohistochemistry of anagen HF in humans shows a paucity of natural killer (NK) cells and γδ T cells around the HF, but a prominence of Langerhans cells, macrophages, mast cells and mature T and B cells (Christoph et al., 2000). The relevance of most of these immune cell types with regard to the HF stem cell niche and immune privilege has only recently been interrogated, and will be discussed in this chapter.

Epidermal stem cells are capable of regenerating HFs in large experimental wounds (Ito et al., 2007), likely an evolutionary adaptation for replacing tissue injury in the wild and the maintenance of protective fur. This *de novo* regeneration of HFs in a wound bed involves TNF signals from macrophages (Wang et al., 2017), FGF9 produced by infiltrating γδ T cells (Gay et al., 2013), as well as interleukin (IL)-1α in the wounding regeneration microenvironment (Yang et al., 2023).

2.3 Melanocytes

The skin also houses melanocyte stem cells in both the IFE and HF. These stem cells are derived from the neural crest, and migrate into the epidermis during embryogenesis. The IFE melanocyte stem cells give rise to melanocytes that transfer melanin to IFE keratinocytes to control skin pigmentation, while the melanocyte stem cells in the HF reside in the lower bulge, and give rise to melanocytes in the HF matrix that ensure hair shaft pigmentation. The keratinocytes of the lower bulge also provide a nice for the melanocyte stem cells (Tanimura et al., 2011). The cyclical activation of the HF keratinocytes *via* paracrine signals from bulge and DP are coupled with melanocyte stem cell activation in order to ensure constant hair pigmentation (Rabbani et al., 2011). Age-related loss of the melanocyte stem cell niche results in pigment loss in the skin, resulting as idiopathic guttate hypomelanosis ("white spots" in aged skin). The melanocyte stem cells in the hair follicle stem cell niche have recently been shown to give rise to progeny that de-differentiate and return to the niche with each hair cycle, and defects in this migration lead to their exhaustion and subsequent hair graying (Sun et al., 2023).

2.4 Sweat glands

Mammalian skin contains eccrine sweat glands that open onto the interfollicular epidermis, and produce sweat that aids in thermoregulation. Stem cells from the sweat gland arise from multipotent epidermal progenitors, and are unipotent, giving rise to the myoepithelial (Krt14+ basal layer) and luminal cells (Krt14^{lo} Krt18+ suprabasal layer) that make up the eccrine sweat duct and gland (Lu et al., 2012), and also contribute to IFE wound healing under certain conditions (Rittie et al., 2013). The eccrine sweat gland niche is likely composed of specialized ECM (Huang et al., 2016), with possible contributions from bone marrow-derived mesenchymal stem cells (Sheng et al., 2009). Sebaceous glands (SGs) that are associated with most HFs are derived from the Lrig1+ epidermal progenitors in the HF junctional zone (Frances & Niemann, 2012; Jensen et al., 2009), and express Blimp1 as they become committed to a SG fate (Sellheyer & Krahl, 2010). Apocrine glands are holocrine glands that are only associated with HFs in areas of secondary sexual maturation (i.e., axillae, groin, nipples), and produce a lipid-rich compound that attains a characteristic odor when processed by skin microbial commensals. The stem cells of the apocrine gland remain unexplored. All sweat glands are richly innervated by sympathetic autonomic nerves, and have distinct ECM and stromal interactions for their niche.

2.5 Merkel cells

The epidermis also contains specialized structures known as touch domes that contain touch-sensitive neuroendocrine Merkel cells. Touch domes are richly innervated by somatic afferents from Aβ nerve fibers (Maricich et al., 2009). Merkel cells are derived from epidermal Krt17+ stem cells which are non-keratinized, are long-lived in mammalian skin, and are the main cell type regulating the Merkel cell niche (Doucet et al., 2013; Van Keymeulen et al., 2009). Merkel cell carcinoma, though extremely rare, is encountered more frequently in patients with systemic immunosuppression (Clarke et al., 2015; Engels et al., 2002), suggesting that immune surveillance of the touch domes is essential for Merkel cell homeostasis.

3. Immune cells and their interactions with the niche

Cells of the immune system have been recognized as crucial players in the function and homeostasis of many tissues and organs. Their ability to respond to various tissue-specific chemokines allow them to home specifically to certain target tissues, and their capacity for migration within and between tissues allows them to mediate long-range cell–cell communication. These distinct properties of the immune system have allowed them to take on non-immune, non-canonical functions in the skin and other organs, in particular to support tissue homeostasis.

The skin, like the gut, provides an interface with the microbe-rich environment and thus has an important immunological role (Abdallah, Mijouin, & Pichon, 2017). Immune cells in the skin are also poised to participate in wound healing and epidermal regeneration. Many immune cells are resident in the epidermis and dermis, but there is also a constant flux of cells to and from the bloodstream. Many cell types are specifically skin-homing, and respond to skin-specific chemokines (Butcher & Picker, 1996; Schaerli et al., 2004).

The immune system is arbitrarily divided into innate and adaptive arms. The innate immune system consists of cells that are generally defined by the lack of specificity of their pattern recognition receptors (PRR) which recognize conserved microbial motifs, and respond to non-specific signals from tissue injury and pathogenic insults. The innate immune system is usually recruited early to sites of injury and include neutrophils, macrophages, dendritic cells and natural killer T cells. Conversely, the adaptive immune system consists of cells that have undergone germline modifications to increase the specificity of their pathogen receptors, and include mature T and B cells (including helper, cytotoxic, regulatory and memory subtypes). The activity and cytokines produced by the innate immune system during the acute phase of inflammation have been shown to influence the fate of adaptive immune cells (Iwasaki & Medzhitov, 2015). During tissue homeostasis, tissue resident immune cells include many cells of the innate immune system, as well as memory cells of the adaptive immune system. These cells have recently been shown to take part in tissue homeostasis in many unexpected ways.

3.1 Innate immune system

As the main sentinels of the immune system, many innate immune cells are positioned in peripheral tissues as a first line of defense against physiological and environmental insults. They play an essential role in wound healing by producing cytokines that support epithelial and mesenchymal cells in repairing tissue injury by activating stem cell populations, as well as fending off pathogenic intruders to maintain a relatively sterile wound bed (MacLeod & Mansbridge, 2016; Portou et al., 2015). The interactions between innate immune cells and various tissue cell types (particularly stem cells) are also present during the steady-state as part of routine tissue homeostasis. For example, invariant natural killer (iNK)T-cells were found to be developmentally programmed to home to skin in neonatal mice where they were central in ideal bacterial commensal colonization, which also had effects on tissue homeostasis and hair follicle development (Wang et al., 2023).

3.1.1 Langerhans cells

Langerhans cells are abundant in the epidermis and express Langerin (CD207). They are seeded in the epidermis during embryogenesis, and their primary function is sampling the epidermis for potential exogenous infiltration and present antigen to T cells in the skin-associated lymphoid tissue (SALT) or draining lymph nodes. Single-cell RNA sequencing has identified at least four subsets of Langerhans cells, with distinct development trajectories and homeostatic roles (Liu et al., 2021). Immunohistochemistry has shown them to be closely apposed to the basal keratinocyte stem cells of the IFE (Parkinson, 1992), as well as Merkel cells (Taira et al., 2002). Their specific function in these stem cell niches are currently unknown.

3.1.2 Dendritic epidermal T cells

In murine skin, dendritic epidermal γδ T cells (DETCs) play a role in immune surveillance *via* their restricted T cell receptor (TCR) devoid of functional diversity (Tigelaar & Lewis, 1995), and are present in large numbers. DETCs support the IFE niche *via* insulin growth factor (IGF)-1 (Sharp et al., 2005), and also contribute to wound healing by producing fibroblast growth factor (FGF)-7 (Jameson et al., 2004). In aging mouse skin, while the epidermal stem cell compartment is relatively stable, the rate of proliferation decreases, and is associated with a reduction in DETC numbers in the epidermis (Giangreco et al., 2008). An increase in HF-associated DETCs is also observed with anagen progression in mice (Hashizume, Tokura, & Takigawa, 1994; Paus et al., 1994), and may play a role in HF stem cell activation.

Similar γδ T cells may be present at much lower levels in the human epidermis (with the majority expressing the αβ TCR), and have been implicated in contact hypersensitivity and wound healing (Toulon et al., 2009), possibly through secreted exosomes (Liu et al., 2022a). Whether epidermal T cells modulate stem cell niches in human skin remains unexplored.

3.1.3 Neutrophils

Neutrophils are the most abundant immune cell with a very short half-life, and are recruited to sites of acute injury to provide the first line of defense against extrinsic pathogens (Mocsai, 2013). Neutrophils also support the peripheral function of B cells, NK cells and T cells (Jaeger et al., 2012; Pillay et al., 2012; Puga et al., 2011). Degranulation of neutrophils results in the release of acute-phase cytokines, chemokines, proteases and reactive oxygen species into the tissue microenvironment—all of which contribute to the early inflammatory response but also sets up the target tissue for wound healing and resolution, and may in turn influence the stem cell niche to regulate stem cell activity during times of stress (Borregaard, Sorensen, & Theilgaard-Monch, 2007; Hager, Cowland, & Borregaard, 2010). In the bone marrow, aged neutrophils migrating back to the stem cell niche is coupled to the circadian release of hematopoietic progenitors into the blood stream (Casanova-Acebes et al., 2013; Mendez-Ferrer et al., 2008). This process is dependent on CXCR4 expression on the neutrophils (Martin et al., 2003).

3.1.4 Macrophages

Myeloid cells like macrophages and other antigen-presenting cells have phagocytic functions central to their immune function as tissue sentinels and surveillance of exogenous antigen, but are also employed clearance of cellular debris during wound healing (DiPietro, 1995).

Tissue macrophages are diverse and highly heterogeneous (Qian et al., 2019), and may be involved in multiple ways to maintain homeostasis. Due to this heterogeneity, categorization of tissue macrophages is difficult, and many distinct subsets have been identified through recent advances of genetic profiling at a single-cell level (Guilliams, Mildner, & Yona, 2018; Zhao et al., 2018). Tissue macrophages can be broadly dividing into the long-lived tissue resident macrophages which originate from the embryonic yolk sac or fetal liver and seeded into tissues *in utero*, and the bone-marrow derived macrophages which are derived from circulating monocytes in adulthood *via* tissue chemotactic factors (Ginhoux & Guilliams, 2016). Most macrophages in the skin are bone marrow derived, unlike other organs such as the brain where microglia are mostly long-lived residents derived from the embryonic yolk sac (Epelman, Lavine, & Randolph, 2014).

Macrophages have also been classified according to their function, namely, the "classical" or "inflammatory" macrophages (previously called M1) that produce interferon, TNF-α and nitric oxide synthase (iNOS) in response to pathogen, and the "alternate" or "anti-inflammatory" (M2) macrophages that promote tissue repair by producing IL-4, IL-13, IL-10 and ornithine/polyamines (Wang, Liang, & Zen, 2014). Detailed analyses of macrophage subsets have shown that this classification may be overly simplistic, and macrophages are tremendously diverse in phenotype and function (Knudsen & Lee, 2016; Martinez & Gordon, 2014). While it is preferred to discuss macrophages according to their precise function, the terms "M1-like" and "M2-like" have been used as shorthand to orientate the reader.

Macrophages are associated with the HF at almost every stage of the hair cycle, with different subsets carrying out specific functions to influence the stem cell niche (Muneeb, Hardman, & Paus, 2019). For example, macrophages increase in numbers in telogen (rest) skin in the mouse, and their depletion is associated with anagen (hair growth) (Castellana, Paus, & Perez-Moreno, 2014). Recent single-cell analyses coupled with functional analysis uncovered a TREM2+ subset of "trichophages" that maintain HF stem cell quiescence *via* Oncostatin M (OSM) during telogen (Wang et al., 2019). Conversely, macrophages (likely other subsets) are postulated to induce HF stem cell activation and thus anagen in certain conditions such as mechanical stretching of the skin (resulting in microtrauma) (Chu et al., 2019). Other studies have identified subsets of macrophages that may induce anagen *via* FGF-5 (Suzuki et al., 1998) or release of Wnt signals during the telogen-to-anagen transition (Castellana et al., 2014). Interestingly, macrophage-derived extracellular vesicles containing Wnt3a and Wnt7b were able to induce anagen in *Balb/c* mice (Rajendran et al., 2020). Further, macrophages also clear cellular debris from HF regression during catagen (Eichmuller et al., 1998; Parakkal, 1969). It is very likely that different subsets and activation states of macrophages are responsible for the control of the HF stem cell niches at various stages of the hair cycle (Amberg et al., 2016). "M1-like" macrophages have been associated with catagen, while "M2-like" macrophages have been postulated to maintain tissue homeostasis in telogen and anagen. The switch between these polarized states has been shown to be affected by intrinsic reactive oxygen species (ROS) associated with apoptotic hair follicles (Liu et al., 2022b). TNF-producing "M1-like" macrophages are implicated in *de novo* hair regeneration in wounds (Wang et al., 2017), as well as anagen initiation after plucking (Chen et al., 2015), while "M2-like" anti-inflammatory tissue resident macrophages are involved in stem cell quiescence during telogen (Wang et al., 2019). future studies which reveal more specific markers for these subsets will aid in understanding their roles more clearly.

3.1.5 Mast cells

Mast cells typically mediate IgE allergic inflammatory responses, and have been shown to be essential in some models of wound healing (Komi, Khomtchouk, & Santa Maria, 2019). Mast cells are also found surrounding HFs and sebaceous glands. IL-6 and TNF-α produced by mast cells have been postulated to influence the innervation of sebocytes, which in turn affects the local substance P concentration of the SG which modulates the proliferation and differentiation of SG stem cells (Toyoda, Nakamura, & Morohashi, 2002).

3.1.6 *Other innate immune cells*

Eosinophils, basophils, innate lymphoid cells (ILCs) and NK cells are other members of the innate immune system. Innate lymphoid cells (ILCs) are derived from the common lymphoid progenitor but do not express mature T/B/NKT markers or antigen receptors, and are involved in wound healing and tissue homeostasis in various epithelial tissues (Rak et al., 2016). Specific interactions between the other cell types and stem cells have not been closely interrogated, and are thus out of the scope of this chapter.

3.2 Adaptive immune system

In contrast to the innate immune system, the hallmarks of the adaptive immune system reflect its mutability and potential for changing in response to environmental or self-antigens. Prime examples are mature T and B cells which produce polyclonal receptors and antibodies that are edited down to match antigen presented to the organism by way of infection, vaccination or mutation (in the case of cancer). The adaptive immune system is also capable of memory, and retains T and B cells with high-affinity receptors which are able to respond in a more efficient and specific manner to re-exposure to the same antigen. Memory cells reside in peripheral lymphoid and somatic tissue, and have recently been shown to participate in tissue homeostasis. Skin-homing lymphocytes express CLA (cutaneous lymphocyte antigen) (Fuhlbrigge et al., 1997), and are also involved in establishing skin-associated lymphoid tissue (SALT), which also involve mast cells and cutaneous nerves (Streilein, Alard, & Niizeki, 1999).

3.2.1 *Mature T and B cells*

Depletion of CD4+ and CD8+ T cells with monoclonal antibodies have been shown to impair wound healing (Barbul et al., 1989), as does thymectomy in mice (Davis et al., 2001). However, these experiments were performed before the discovery of innate lymphoid cells, which bear the same CD4 and CD8 markers as mature T cells. It was an open question whether antigen-specific mature T cells contributed to skin homeostasis.

The skin microbiome plays a significant role in "training" the immune system and thus helping to dictate a specific immune vigor and tone. The presence of microbiota in the skin leads to antigen presentation *via* MHC molecules which leads to maturation of T and B cells. Mature CD8+ T cells (Tc17 and Tc1 subsets) have been shown to accelerate wound healing in the skin in a MHC- and commensal-dependent manner (Linehan et al., 2018). This suggests that the experience of the immune system plays a role in tissue homeostasis. Similarly, in other systems, immunological age (i.e., the balance between naïve/mature lymphocytes) influences the degree of proliferation and differentiation of osteogenic precursors after fractures, and had a significant impact on the rate of bone healing and remodeling (Bucher et al., 2019).

3.2.2 Regulatory T cells

CD4+ CD25+ FoxP3+ T regulatory (T reg) cells patrol the peripheral immune system to prevent autoimmunity and maintain self-tolerance (Fehervari & Sakaguchi, 2004). These processes are essential for stem cell niches, which are typically sites of immune privilege (Niederkorn & Stein-Streilein, 2010). T regs express cell surface and secretory factors that modulate the activity of other cells in its vicinity, and are thus poised to be important players in the stem cell niche of many tissues (Lane et al., 2014; Taams et al., 2005). T regs have been found to be in close association with the hematopoietic stem cell (HSC) niche in the bone marrow (Fujisaki et al., 2011), and are known to help maintain the donor stem cells pool after bone marrow transplantation (Edinger et al., 2003). In the skin, resident T regs (which represent about 20% of resident T cells) migrate to the skin during HF development and the acquisition of the skin microbiome in a CCL20-CCR6 dependent manner (Scharschmidt et al., 2017). T reg impairment in Scurfy mice results in profound autoimmunity and skin inflammation and alopecia (Yang et al., 2015). After plucking, resident T regs may mediate the subsequent induced anagen after hair plucking *via* Jagged1–Notch interactions with the HF stem cell niche (Ali et al., 2017). Glucocorticoid receptor signaling in T regs also appears essential for the maintenance of the HF stem cell niche and normal progression of the hair cycle (Liu et al., 2022c), which might be a mechanism through which stress-induced hair loss (i.e., Telogen effluvium) occurs.

4. Immune dysregulation and stem cells

Dysfunction of the stem cell niche results in the impairment of essential stem cell properties. For example, loss of attachment to vital basement membrane ECM proteins affects the proliferative state of stem cells. Dysregulation or exhaustion of stem cells in a compromised niche ultimately leads to impairment in wound healing and tissue homeostasis, senescence and loss of skin and hair integrity. Collapse of immune privilege of stem cell niches also lead to premature depletion of crucial stem cell populations.

Epidermal stem cells are sensitive to changes in their microenvironment, and cytokines produced by immune cells can influence the balance between proliferative or quiescence signals in the stem cell niche.

4.1 Psoriasis

Psoriasis provides an intriguing model to study immune cell–niche interactions in the IFE. IL-17 and IL-22 produced by Th17 cells are believed to drive epidermal hyperproliferation in psoriatic lesions (van der Fits et al., 2009). The same cytokines have also been shown to increase the expression of stemness markers in the basal layer of keratinocytes, which is expanded to include the suprabasal Krt10 layer in psoriasis (Ekman et al., 2019).

4.2 Cancer stem cells

While the stem cell compartments of the skin enjoy relative immune privilege during the steady-state, the same immune cells (antigen-presenting cells, NK T cells and T regs) are responsible for the clearance of cells that have accumulated enough DNA damage as part of their role in immunosurveillance of cancers (Moodycliffe et al., 2000). Disruption in this process, either by hijacking and weaponizing cell autonomous immune privilege tools (i.e., downregulating MHC molecules) or by defects in the immune surveillance mechanism (Browning & Bodmer, 1992; Hahne et al., 1996), allows for "immune escape" and proliferation of skin tumors (de Visser, Eichten, & Coussens, 2006).

Established malignant tumors have their own cancer stem cell niches, whose microenvironment is also regulated by immune cells, in particular tumor-associated macrophages (TAMs) (Fujimura et al., 2018; Noy & Pollard, 2014). Squamous cell carcinoma arising from oncogenic mutations in the IFE basal stem cells contains TAMs that support the cancer stem cell niche by secreting vascular endothelial growth factor (VEGF) (Moussai et al., 2011). The inflammatory/anti-inflammatory balance (i.e., "M1-like" vs "M2-like" skew) of TAMs in melanoma is also associated with survival and prognosis, with "M2-like" anti-inflammatory TAMs having a poorer prognosis—possibly by enabling immune evasion (Falleni et al., 2017).

In addition to TAMs, mast cells are also implicated in stimulating angiogenesis in a wide variety of cancers (Ribatti & Crivellato, 2012). Mast cells are commonly found within a variety of skin tumors (Cawley & Hoch-Ligeti, 1961), and a higher numbers of mast cells is associated with a poorer prognosis in Merkel cell carcinoma (Beer, Ng, & Murray, 2008).

4.3 Autoimmunity

Conversely, the collapse of immune privilege results in autoimmune attack of stem cells in their niche by cytotoxic NKT cells (Ito et al., 2008). Autoimmune destruction of IFE melanocyte stem cells results in vitiligo (Rork, Rashighi, & Harris, 2016). In the HF, depending on which stem cell population is affected, this can results in non-scarring alopecia areata if the attack occurs around the bulb or matrix keratinocytes (Paus & Bertolini, 2013), or scarring alopecia (i.e., lichen planopilaris or discoid lupus) is the attack is directed to the HF bulge region (Harries et al., 2013). The mechanism of immune privilege is believed to be the relative expression of MHC molecules on the stem cells, which is in turn modulated by niche signals. Whether anti-inflammatory "M2-like" macrophages or T reg cells in the skin are involved in the establishment of physiological immune privilege, as they do in pathological tumors (Zhao et al., 2012), remains an open question.

4.4 Aging and senescence

Age-related physiological changes in the skin and hair are a result of intrinsic aging of stem cells (e.g., waning DNA repair and replication infidelity), coupled with extrinsic aging mediated by environmental insults such a accumulation of UV mutations and oxidative damage. This results in a weakened skin barrier, increased transepidermal water loss, poorer wound healing, wrinkles, pigmentary dyschromia of skin and hair, hair thinning, and increased risk of malignancies (Mimeault & Batra, 2010). While much research has focused on the senescence of epidermal and mesenchymal cells in the skin, the role of the senescent immune system has not been explored in the maintenance of epidermal stem cell niches. Aging is associated with a reduction in Langerhans cells and DETCs (Giangreco et al., 2008; Sunderkotter, Kalden, & Luger, 1997), which provide immune surveillance of the epidermal stem cells and participate in the clearance of potentially pathogenic or deleterious mutations. Thus, impairment of immune–niche interactions with age should gain more scientific attention as the global population ages and lifespans continue to lengthen.

5. Studying immune-stem cell interactions

The interactions between immune cells and the epidermal stem cell niche have been alluded to for decades, with exhaustive descriptive studies associating the distribution of various immune cells in the epidermis and associated appendages (Christoph et al., 2000). However, the functional relationship between immune cells and the niche remains elusive as sophisticated tools required to dissect their interaction have only been developed in the last few years. As immune cells are migratory and transient in tissues, and usually at a much lower proportions compared with epithelial and stromal cells in unperturbed tissue, their representation on bulk transcriptomic analysis (i.e., RNA seq) may be masked.

5.1 Single-cell transcriptomics

The ability to study RNA expression at the single-cell level has allowed researchers to circumvent the dilution of immune cell signals in bulk RNA-sequencing experiments (Chattopadhyay et al., 2014). Briefly, tissue is enzymatically dissociated into a single-cell suspension, which is then parsed into single cell units labeled with a nucleotide barcode tag (usually with microfluidics or emulsification technology), and the reverse transcriptase PCR reaction is performed on each cell individually (Fig. 2A(i)). The data from each cell is pooled with all the other cells from the tissue and analyzed with algorithms that compare their similarities. This allows these differences to be visualized as clusters. Single-cell RNA-sequencing has also been used to analyze the epithelial and stromal compartments of the skin and HF (Jensen & Watt, 2006; Joost et al., 2016; Yang et al., 2017). By performing single-cell RNA-sequencing on just the immune cells (i.e., by sorting for CD45+ cells in the niche), potentially rare subsets of immune cells can be identified (Nguyen et al., 2018). For examples, a rare subset of TREM2+ macrophages associated with HFs was discovered by this method (Wang et al., 2019). Further validation studies showed that these macrophages produced OSM that inhibited HF stem cell activation, and played a role in maintaining quiescence in the HF niche. Additionally, transition states of immune cell subsets can be identified with this technology, and "pseudotime" or "pseudodifferentiation" analyses can be carried out to generate hypotheses for further validation.

For the bone marrow, the opposite strategy was employed (i.e., analyzing the $CD45^{low}$) cells to studying the hematopoietic niche *sans* the immune cells (Tikhonova et al., 2019). Single-cell RNA-sequencing of entire tumors have also identified the immune composition specific to certain cancers, like melanoma (Tirosh et al., 2016), but other experiments are still required to determine their significance *in vivo*.

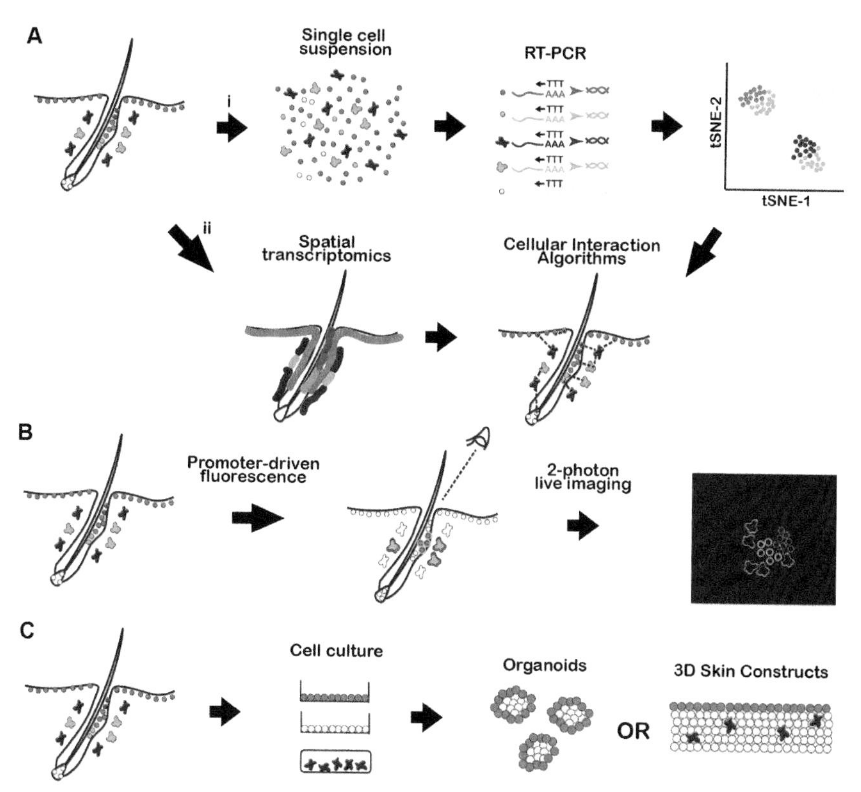

Fig. 2 Technologies for studying niche interactions. Representative cartoons of modern methods that are being employed in studying the interaction between immune cells and the stem cell niche. (A) (i) Single-cell RNA sequencing. Tissue is dissociated mechanically and enzymatically into a single-cell suspension. Single cells are then individually analyzed with a RT-PCR reaction to create barcoded cDNA libraries which are sequenced in a high-throughput manner. The gene expression profile of each cell is compared with all other cells, these differences are represented as Euclidean distances on a tSNE (t-distributed stochastic neighbor embedding) plot. In this example, epidermal stem cells (orange and green circles) resemble each other more than they do the immune cells (dark and light blue) which cluster separately. Dermal cells (yellow) are represented by yet another cluster. Increased resolution of immune cells can be achieved by selecting immune cells by FACS, so as to remove the noise from epidermal and stromal cells. This method has also been used for single-cell genomic, proteomic and epigenomic studies. Cellular interaction algorithms like CellChat and CellPhoneDB can then be used to match ligand–receptor interactions on cell clusters to predict potential interactions within the tissue. (ii) Spatial transcriptomics. Tissue mRNAs are hybridized *in situ,* maintaining spatial information, and sequenced on a chip that allows researchers to isolate and pinpoint gene expression of cells with great resolution. Together with cellular interaction algorithms, these methods will prove to be very useful in understanding tissue microenvironments and stem cell niches. (B) Intravital fluorescence imaging. Stem cells and immune cells can be labeled with fluorescent reporter

(Continued)

Cellular Interaction Algorithms Single-cell RNA sequencing, while able to detail cellular diversity and heterogeneity, does not give us insight into the complex interactions between individual cells. Cellular interaction algorithms have been developed to infer cellular interactions from single-cell sequencing data. These algorithms can reveal the cellular networks and communication pathways involved during the human hair cycle, particularly between the immune cells and the epithelial/mesenchymal compartments of the hair follicle.

Cellular interaction algorithms use statistical methods and machine learning approaches to analyze single-cell sequencing data to identify potential interactions between cells (Fig. 2A). For example, CellPhoneDB is a widely used tool that identifies cell-to-cell interactions by comparing the expression of ligands and receptors between different cell types (Efremova et al., 2020). Other algorithms, such as SCENIC and CellChat, use gene regulatory networks and signaling pathways to infer cellular interactions, taking into account not only known receptor-ligand pairs, but co-receptors and signaling apparatus as well (Jin et al., 2021). However, while these algorithms may lend some insight into cellular interactions, this analysis needs to be interpreted carefully as the output is biased toward the known interactions in the input databases, and does not given any spatial information.

Fig. 2—Cont'd proteins using a promoter driven approach. These cells will fluorescence under a 2-photon microscopy and their behavior can be observed *in situ*. In this figure, stem cells are labeled with a green fluorescence, while immune cells have a red fluorescence. However, this technique is limited to experimental model animals, and these animals need to be anesthetized and immobilized for up to 16–20h in specialized rigs for the intravital images to be acquired. Results from these experiments allow us to observe the interactions, motility, and dynamics of cells in their native niche, and are mesmerizing and fascinating when presented at scientific meetings. (C) Organoid culture and 3D skin constructs. Sophisticated culture techniques and intricate engineering platforms have pushed the limitations of two-dimensional monolayer culture of stem cells. Self-assembly of several cell types into organoids, or arranging cells into three-dimensional constructs have allowed researchers to try to recreate the stem cell niche *ex vivo*. These platforms allow us to manipulate niche dynamics, test drugs on the system, and hopefully at some point replace animal testing. Induced pluripotent stem cells have also been used to generate different tissue types for organoids/skin constructs, and may lead to a more non-invasive way of utilizing patient cells. Addition of immune cells into these systems will be the next step to dissecting their relationships and interactions with stem cell niches.

5.2 Spatial transcriptomics

Spatial transcriptomics is a cutting-edge technique that enables the profiling of gene expression in a tissue section with spatial information (Fig. 2A(ii)) (Stahl et al., 2016). This technology allows for the visualization of gene expression patterns in their native tissue context, which can provide a deeper understanding of the spatial organization of cells and their interactions. By integrating spatial transcriptomics with single-cell sequencing data, researchers can gain a more comprehensive understanding of the molecular and cellular mechanisms underlying complex biological processes.

One of the strengths of spatial transcriptomics is its ability to capture the spatial heterogeneity of gene expression within tissues. This can reveal important insights into cellular differentiation, migration, and communication that would not be apparent from single-cell sequencing data alone. For example, a recent study used spatial transcriptomics to map the transcriptome of a breast tumor and identified distinct subpopulations of cells with specific gene expression profiles, some of which were spatially localized to specific regions of the tumor. This study demonstrated the power of spatial transcriptomics to uncover previously hidden cellular heterogeneity and may have important implications for the diagnosis and treatment of breast cancer.

5.3 *In vivo* live imaging

Two-photon live imaging has revolutionized our perspective on cellular behavior *in vivo* and *in vitro*. Fluorescent labeling of live cells (usually *via* H2 histone promotor—green fluorescent protein (GFP) expression) coupled with sophisticated rigs has allowed researchers to visualize cellular interactions and motility in culture, as well as *in situ* in a living organism (Fig. 2B). The skin has the advantage of being easily accessible, and requires an immobilized, anesthetized animal with a window chamber. Using this technology, epidermal stem cell competition and behavior within their niche have been observed for the very first time (Rompolas & Greco, 2014; Rompolas et al., 2013). Other cells of the stem cell niche can be similarly labeled with more complex combinations of fluorescent reporter proteins coupled to the appropriate promoters, and this has been used to visualize immune cells (Pittet et al., 2018), endothelial cells and stromal cells during the steady-state, and during tissue injury. Combining this technology with multiphoton or multispectral imaging will allow visualization of more than 2 cell types at once.

Differential labeling of macrophages and neutrophils revealed that "M2-like" resident macrophages prevents neutrophil-led inflammation during microtrauma caused by single-cell laser ablation (Uderhardt et al., 2019). This is a form of immune privilege that is conferred to tissues that may undergo frequent remodeling so as to prevent unnecessary immune attention. Using this technology to study macrophage dynamics during the hair cycle will yield fascinating insight into the immune–niche interactions in this system. However, this technology is currently limited to experimental animal models which are amenable to genetic fluorescence manipulation.

5.4 Organoid culture and 3D constructs

By placing a stem cell type in two-dimensional culture, it is removed from its niche and surrounding interactions (Fig. 2C). Moreover, the nature of its attachment is changed from a complex ECM, which promotes stemness (Trappmann et al., 2012), to at best coated polystyrene with one or two ECM components (Reichelt, 2007). This artificial *in vitro* system, while convenient, changes the epigenetics and phenotype of the stem cell so that it does not truly represent the *in vivo* version of itself. Both keratinocytes (Vollmers et al., 2012) and DP cells undergo drastic transcriptomic shifts when placed in two-dimensional culture, losing their stemness and hair inductive abilities, and this is partially recovered by three-dimensional spheroid culture. Co-cultures of keratinocytes and fibroblasts provide a better model to study wound healing (Wojtowicz et al., 2014), but are currently still do not incorporate immune cells so as to be more representative. While *ex vivo* techniques like whole skin explant culture (Frade et al., 2015) and HF organ culture (i.e., the Philpott assay (Philpott, 2018)) may shed some light on tissue interactions, these tissues are hard to obtain and may have patient- and site-specific variations that are hard to control for robust research.

Attempts to re-create the stem cell niche in organoid culture in a controlled manner, whereby multiple cell types from the native tissue are allowed to aggregate and replicate with *in vivo* relationships, has been pioneered in the study of intestinal stem cells (Date & Sato, 2015). Skin and hair organoids have also been utilized to study the skin niche (Lee et al., 2018), but the hierarchical and complex nature of stem cell compartments in the skin and hair and other appendages may require more complex and bio-engineered culture systems to truly study them *ex vivo*. Efforts to incorporate multiple cell types into a skin construct has recently been achieved (Abaci et al., 2015, 2016, 2018; Gledhill et al., 2015). The current limitation in this field is the inability to recreate and model the hair cycle *ex vivo*, possibly due to the loss of specific niche signals from immune cells. The study of complex tissue interactions is being tackled now with organ-on-a-chip techniques, using engineering expertise of microfluidics and biosensors to fabricate an *ex vivo* system amenable for experimental manipulation (Liu et al., 2023).

5.5 Digital organoids and other *in silico* methods

The exponential advances in computing power will soon allow us to combine data from all previously discussed methods into a workable *in silico* model to study the human hair cycle. The human hair cycle can last up to 10 years, which makes *in vivo* observations and perturbations extremely challenging. The multiple stem cell compartments and cell types involved in maintaining the hair cycle also limit the accuracy of *in vitro* and *ex vivo* studies. Using state-of-the-art computer models, coupled with artificial intelligence and algorithms, we will be able to visualize the hair cycle *in silico* and subject it to perturbations such as stress, illness, mutations and medications. Digital organoids have already been used to study adult human retinal tissue (Wahle et al., 2023), and future iterations of this technology will be very useful in the study of the human hair cycle.

6. Harnessing the immune system to for hair regrowth

With increasing understanding of the interactions between the immune cells and the hair follicle, we will be able to develop targeting immunotherapy for hair regrowth and regeneration will be possible. Current therapies that harness the immune system for this function have been shown to be safe and effective for non-scarring alopecias.

6.1 Contact immunotherapy

In alopecia areata (AA), in which hair loss results from autoreactive NKG2D+ T cells attacking the anagen hair follicle, contact immunotherapy (CIT) with sensitizers like squaric acid dibutylester (SADBE) and diphencyprone (DPCP) is an established option for patients who are not suitable for long-term immunosuppression. Eligible patients are sensitized at the start of therapy, and thereafter receive a weekly topical application of the sensitizer to the scalp, at increasing concentrations according to the level of allergic dermatitis and itch that results. This usually results in 60–80% response rate by 4–6 months, but requires a significant time commitment and tolerance for messy treatments. The precise mechanism of CIT is unknown, and is believed to work by modulating the scalp immune milieu from a predominantly Th1/Th17 in AA to a Th2 skew in remission.

Patients undergoing CIT represent an ideal system to study the mechanism of CIT. The change in the immune polarization around the HF stem cell niche during CIT can be characterized with approaches listed above, and analyzed according to responders and non-responders, to identify the key pathways associated with AA remission. This will lead to the discovery of new targets for AA treatment, which can be formulated into a more convenient treatment modality than CIT itself.

6.2 Microneedling

Microneedles have emerged as a promising technology in the field of dermatology, including their application in the treatment of hair loss. Microneedling involves the use of tiny needles ranging from 0.3 to 2.0 mm to create controlled microtrauma in the skin. These microneedles penetrate the scalp at a predetermined depth, stimulating the production of growth factors and activating the scalp's wounding response. Several studies have investigated the efficacy of microneedling in treating androgenetic alopecia. Microneedling combined with minoxidil solution was shown to result in significant hair regrowth in patients with androgenetic alopecia in a split-scalp study (Dhurat et al., 2013).

Microneedling can be delivered in many ways: from dermarollers that are available on most online shopping platforms, microneedle patches that can be loaded with drugs and biological actives, or laser-assisted devices that create microchannels in the scalp. Again, the exact mechanism by which the wounding response leads to hair regrowth in humans is unknown. In mouse models, it is widely accepted that any trauma to the telogen back fur results in anagen entry, and is not analogous to the human scalp. Newer models, likely *ex vivo* or even *in silico*, are essential to understanding the effects of microtrauma on the HF stem cell niche.

6.3 Platelet-rich plasma (PRP)

PRP has also emerged as a promising adjunct therapy for all forms of non-scarring alopecia. PRP is an autologous product derived from the patient's own blood obtained by centrifugation, containing a high concentration of platelets rich in growth factors. When injected intradermally into the scalp, activation of the platelets leads to the release of growth factors, such as platelet-derived growth factor (PDGF), vascular endothelial growth factor (VEGF), and insulin-like growth factor (IGF), among others. These factors promote angiogenesis, increasing flow to the hair follicles, and enhance nutrient delivery to the scalp. These growth factors may also directly stimulate hair follicle stem cells, promoting hair growth and preventing hair miniaturization. Additionally, PRP has been shown to promote the production of extracellular matrix proteins, such as collagen and elastin, which play a crucial role in maintaining the structural integrity of the scalp and supporting hair growth.

Several systematic reviews have concluded that PRP is effective and safe in AGA, with or without conventional medical treatments (Gupta & Bamimore, 2022a, 2022b). A randomized controlled split-scalp study comparing PRP to sham injections also showed that while PRP was effective in improving hair density, saline injections had some positive effects on preventing miniaturization (Chuah et al., 2023), suggesting that both PRP and microneedle-induced microtrauma may have synergistic actions in treating AGA. PRP may also have some immunomodulatory and anti-inflammatory effects, which is reflected in its positive effects in AA in a few small case series (Nouh, Abdelaal, & Fathy, 2023). However, the variations in protocols of PRP preparation, activation and injection leads to a varied landscape of efficacy (Li et al., 2023a), and more studies with stringent protocols are required to identify the optimal settings for the best effects.

6.4 Exosomes

Exosomes, small extracellular vesicles derived from various cell types, may also have a potential role in the treatment skin and hair conditions. Like the platelets in PRP, exosomes contain a diverse range of bioactive molecules, including growth factors, cytokines, and nucleic acids. Exosomes can be derived from stem cells or immune cells, and also bear cell-specific membranes that have bioactive receptors that can interact with target cell types. Exosomes also contain small interfering RNAs (siRNAs) or microRNAs (miRNAs), which can modulate gene expression and target specific pathways involved in hair growth regulation.

For example, exosomes from cultured dermal papilla cells have been shown to modulate HF stem cell activity *via* Wnt-LEF1 signaling (Li et al., 2023b). Mesenchymal stem cells in adipose tissue have also been used to develop exosomes containing the microRNA miR-122-5p that is able to blunt the effect of dihydrotestosterone (DHT) in cultured dermal papilla cells and C57BL mice (Liang et al., 2023). Large-scale production of "engineered nanovesicles" derived from fibroblasts and macrophages has been attempted to capture the benefits of exosomes (Gangadaran et al., 2022; Rajendran et al., 2021), and these products appear to have positive effects on the hair cycle in C57BL mice as well.

This approach holds promise for personalized and precision medicine in the treatment of hair loss. However, further research is needed to optimize the isolation methods, cargo loading, and delivery strategies of exosomes for effective hair regeneration.

6.5 Laser-assisted microtrauma

Laser-assisted microtrauma is another method that has been deployed to treat non-scarring alopecias. Photobiomodulation with low-level laser therapy (LLLT) is an FDA-cleared home device that utilizes red or near-infrared light to stimulate cellular activity in the hair follicles. The laser energy is absorbed by the mitochondria of the cells, promoting increased production of adenosine triphosphate (ATP) and enhancing cellular metabolism. This, in turn, can lead to improved hair follicle function and hair growth. While several studies have explored the efficacy of LLLT in treating hair loss conditions such as androgenetic alopecia, the real-life efficacy has been modest to disappointing at best (Qiu et al., 2022).

Fractional ablative and non-ablative lasers have also been used in both AGA and AA, alone or in combination with medical treatment or PRP (Bertin, Vilarinho, & Junqueira, 2018; Day, McCarthy, & Talaber, 2022). These lasers, like microneedling, create columns of controlled microtrauma in the scalp, stimulating local wounding responses that appear to be beneficial in promoting hair regrowth and regeneration. Fractional radiofrequency microneedling is another modality that may be useful for AGA (Tan et al., 2019). However, these methods have not been tested vigorously, and further randomized controlled trials and meta-analyses are required to confirm their efficacy.

7. Conclusion

The contributions of immune cells to the regulation and maintenance of stem cell niches are beginning to be appreciated. Not just in the skin, but also in the bone marrow, gut, skeletal muscle and lungs. It is likely that both innate immune cells, and resident memory cells of the adaptive immune system, are intimately involved in most stem cell niches. Understanding these interactions will allow us to predict adverse effects of immunosuppressants or small molecule drugs on tissue homeostasis, harness the immune system in the management of stem cells disorders like cancer and aging, as well as reverse engineer the stem cell niche *ex vivo* to accelerate achievements in regenerative medicine.

References

Abaci, H. E., et al. (2015). Pumpless microfluidic platform for drug testing on human skin equivalents. *Lab on a Chip*, *15*(3), 882–888.

Abaci, H. E., et al. (2016). Human skin constructs with spatially controlled vasculature using primary and iPSC-derived endothelial cells. *Advanced Healthcare Materials*, *5*(14), 1800–1807.

Abaci, H. E., et al. (2018). Tissue engineering of human hair follicles using a biomimetic developmental approach. *Nature Communications*, *9*(1), 5301.

Abdallah, F., Mijouin, L., & Pichon, C. (2017). Skin immune landscape: Inside and outside the organism. *Mediators of Inflammation*, *2017*, 5095293.

Ali, N., et al. (2017). Regulatory T cells in skin facilitate epithelial stem cell differentiation. *Cell*, *169*(6), 1119–1129 e11.

Amberg, N., et al. (2016). Effects of Imiquimod on hair follicle stem cells and hair cycle progression. *The Journal of Investigative Dermatology*, *136*(11), 2140–2149.

Barbul, A., et al. (1989). The effect of in vivo T helper and T suppressor lymphocyte depletion on wound healing. *Annals of Surgery*, *209*(4), 479–483.

Beer, T. W., Ng, L. B., & Murray, K. (2008). Mast cells have prognostic value in Merkel cell carcinoma. *The American Journal of Dermatopathology*, *30*(1), 27–30.

Bertin, A. C. J., Vilarinho, A., & Junqueira, A. L. A. (2018). Fractional non-ablative laser-assisted drug delivery leads to improvement in male and female pattern hair loss. *Journal of Cosmetic and Laser Therapy*, *20*(7–8), 391–394.

Blanpain, C., & Simons, B. D. (2013). Unravelling stem cell dynamics by lineage tracing. *Nature Reviews. Molecular Cell Biology*, *14*(8), 489–502.

Borregaard, N., Sorensen, O. E., & Theilgaard-Monch, K. (2007). Neutrophil granules: A library of innate immunity proteins. *Trends in Immunology*, *28*(8), 340–345.

Braun, K. M., & Prowse, D. M. (2006). Distinct epidermal stem cell compartments are maintained by independent niche microenvironments. *Stem Cell Reviews*, *2*(3), 221–231.

Browning, M. J., & Bodmer, W. F. (1992). MHC antigens and cancer: Implications for T-cell surveillance. *Current Opinion in Immunology*, *4*(5), 613–618.

Bucher, C. H., et al. (2019). Experience in the adaptive immunity impacts bone homeostasis, remodeling, and healing. *Frontiers in Immunology, 10*, 797.
Butcher, E. C., & Picker, L. J. (1996). Lymphocyte homing and homeostasis. *Science, 272*(5258), 60–66.
Casanova-Acebes, M., et al. (2013). Rhythmic modulation of the hematopoietic niche through neutrophil clearance. *Cell, 153*(5), 1025–1035.
Castellana, D., Paus, R., & Perez-Moreno, M. (2014). Macrophages contribute to the cyclic activation of adult hair follicle stem cells. *PLoS Biology, 12*(12), e1002002.
Cawley, E. P., & Hoch-Ligeti, C. (1961). Association of tissue mast cells and skin tumors. *Archives of Dermatology, 83*, 92–96.
Chattopadhyay, P. K., et al. (2014). Single-cell technologies for monitoring immune systems. *Nature Immunology, 15*(2), 128–135.
Chen, C. C., et al. (2015). Organ-level quorum sensing directs regeneration in hair stem cell populations. *Cell, 161*(2), 277–290.
Christoph, T., et al. (2000). The human hair follicle immune system: Cellular composition and immune privilege. *The British Journal of Dermatology, 142*(5), 862–873.
Chu, S. Y., et al. (2019). Mechanical stretch induces hair regeneration through the alternative activation of macrophages. *Nature Communications, 10*(1), 1524.
Chuah, S. Y., et al. (2023). Efficacy of platelet-rich plasma in Asians with androgenetic alopecia: A randomized controlled trial. *Indian Journal of Dermatology, Venereology and Leprology, 89*(1), 135–138.
Clarke, C. A., et al. (2015). Risk of merkel cell carcinoma after solid organ transplantation. *Journal of the National Cancer Institute, 107*(2).
Date, S., & Sato, T. (2015). Mini-gut organoids: Reconstitution of the stem cell niche. *Annual Review of Cell and Developmental Biology, 31*, 269–289.
Davis, P. A., et al. (2001). Effect of CD4(+) and CD8(+) cell depletion on wound healing. *The British Journal of Surgery, 88*(2), 298–304.
Day, D., McCarthy, M., & Talaber, I. (2022). Non-ablative Er:YAG laser is an effective tool in the treatment arsenal of androgenetic alopecia. *Journal of Cosmetic Dermatology, 21*(5), 2056–2063.
de Visser, K. E., Eichten, A., & Coussens, L. M. (2006). Paradoxical roles of the immune system during cancer development. *Nature Reviews. Cancer, 6*(1), 24–37.
Dhurat, R., et al. (2013). A randomized evaluator blinded study of effect of microneedling in androgenetic alopecia: A pilot study. *The International Journal of Trichology, 5*(1), 6–11.
DiPietro, L. A. (1995). Wound healing: The role of the macrophage and other immune cells. *Shock, 4*(4), 233–240.
Doucet, Y. S., et al. (2013). The touch dome defines an epidermal niche specialized for mechanosensory signaling. *Cell Reports, 3*(6), 1759–1765.
Edinger, M., et al. (2003). CD4+CD25+ regulatory T cells preserve graft-versus-tumor activity while inhibiting graft-versus-host disease after bone marrow transplantation. *Nature Medicine, 9*(9), 1144–1150.
Efremova, M., et al. (2020). CellPhoneDB: inferring cell-cell communication from combined expression of multi-subunit ligand-receptor complexes. *Nature Protocols, 15*(4), 1484–1506.
Eichmuller, S., et al. (1998). Clusters of perifollicular macrophages in normal murine skin: Physiological degeneration of selected hair follicles by programmed organ deletion. *The Journal of Histochemistry and Cytochemistry, 46*(3), 361–370.
Ekman, A. K., et al. (2019). IL-17 and IL-22 promote keratinocyte stemness in the germinative compartment in psoriasis. *The Journal of Investigative Dermatology.*
Engels, E. A., et al. (2002). Merkel cell carcinoma and HIV infection. *Lancet, 359*(9305), 497–498.

Epelman, S., Lavine, K. J., & Randolph, G. J. (2014). Origin and functions of tissue macrophages. *Immunity, 41*(1), 21–35.
Falleni, M., et al. (2017). M1 and M2 macrophages' clinicopathological significance in cutaneous melanoma. *Melanoma Research, 27*(3), 200–210.
Fehervari, Z., & Sakaguchi, S. (2004). CD4+ Tregs and immune control. *The Journal of Clinical Investigation, 114*(9), 1209–1217.
Festa, E., et al. (2011). Adipocyte lineage cells contribute to the skin stem cell niche to drive hair cycling. *Cell, 146*(5), 761–771.
Frade, M. A., et al. (2015). Prolonged viability of human organotypic skin explant in culture method (hOSEC). *Anais Brasileiros de Dermatologia, 90*(3), 347–350.
Frances, D., & Niemann, C. (2012). Stem cell dynamics in sebaceous gland morphogenesis in mouse skin. *Developmental Biology, 363*(1), 138–146.
Fuchs, E., Tumbar, T., & Guasch, G. (2004). Socializing with the neighbors: Stem cells and their niche. *Cell, 116*(6), 769–778.
Fuhlbrigge, R. C., et al. (1997). Cutaneous lymphocyte antigen is a specialized form of PSGL-1 expressed on skin-homing T cells. *Nature, 389*(6654), 978–981.
Fujimura, T., et al. (2018). Tumor-associated macrophages: Therapeutic targets for skin cancer. *Frontiers in Oncology, 8*, 3.
Fujisaki, J., et al. (2011). In vivo imaging of Treg cells providing immune privilege to the haematopoietic stem-cell niche. *Nature, 474*(7350), 216–219.
Gangadaran, P., et al. (2022). Application of cell-derived extracellular vesicles and engineered nanovesicles for hair growth: From mechanisms to therapeutics. *Frontiers in Cell and Development Biology, 10*, 963278.
Gay, D., et al. (2013). Fgf9 from dermal gammadelta T cells induces hair follicle neogenesis after wounding. *Nature Medicine, 19*(7), 916–923.
Giangreco, A., et al. (2008). Epidermal stem cells are retained in vivo throughout skin aging. *Aging Cell, 7*(2), 250–259.
Ginhoux, F., & Guilliams, M. (2016). Tissue-resident macrophage ontogeny and homeostasis. *Immunity, 44*(3), 439–449.
Gledhill, K., et al. (2015). Melanin transfer in human 3D skin equivalents generated exclusively from induced pluripotent stem cells. *PLoS One, 10*(8), e0136713.
Goldstein, J., & Horsley, V. (2012). Home sweet home: Skin stem cell niches. *Cellular and Molecular Life Sciences, 69*(15), 2573–2582.
Greco, V., et al. (2009). A two-step mechanism for stem cell activation during hair regeneration. *Cell Stem Cell, 4*(2), 155–169.
Guilliams, M., Mildner, A., & Yona, S. (2018). Developmental and functional heterogeneity of monocytes. *Immunity, 49*(4), 595–613.
Gupta, A. K., & Bamimore, M. (2022a). Platelet-rich plasma monotherapies for androgenetic alopecia: A network meta-analysis and meta-regression study. *Journal of Drugs in Dermatology, 21*(9), 943–952.
Gupta, A. K., & Bamimore, M. A. (2022b). The effect of placebo in split-scalp and whole-head platelet-rich plasma trials for androgenetic alopecia differs: Findings from a systematic review with quantitative evidence syntheses. *Journal of Cosmetic Dermatology, 21*(4), 1454–1463.
Hager, M., Cowland, J. B., & Borregaard, N. (2010). Neutrophil granules in health and disease. *Journal of Internal Medicine, 268*(1), 25–34.
Hahne, M., et al. (1996). Melanoma cell expression of Fas(Apo-1/CD95) ligand: Implications for tumor immune escape. *Science, 274*(5291), 1363–1366.
Halprin, K. M. (1972). Epidermal "turnover time"—A re-examination. *The British Journal of Dermatology, 86*(1), 14–19.
Harries, M. J., et al. (2013). Lichen planopilaris is characterized by immune privilege collapse of the hair follicle's epithelial stem cell niche. *The Journal of Pathology, 231*(2), 236–247.

Hashizume, H., Tokura, Y., & Takigawa, M. (1994). Increased number of dendritic epidermal T cells associated with induced anagen phase of hair cycles. *Journal of Dermatological Science*, *8*(2), 119–124.
Hsu, Y. C., Li, L., & Fuchs, E. (2014). Emerging interactions between skin stem cells and their niches. *Nature Medicine*, *20*(8), 847–856.
Hsu, Y. C., Pasolli, H. A., & Fuchs, E. (2011). Dynamics between stem cells, niche, and progeny in the hair follicle. *Cell*, *144*(1), 92–105.
Huang, S., et al. (2016). 3D bioprinted extracellular matrix mimics facilitate directed differentiation of epithelial progenitors for sweat gland regeneration. *Acta Biomaterialia*, *32*, 170–177.
Ito, M., et al. (2005). Stem cells in the hair follicle bulge contribute to wound repair but not to homeostasis of the epidermis. *Nature Medicine*, *11*(12), 1351–1354.
Ito, M., et al. (2007). Wnt-dependent de novo hair follicle regeneration in adult mouse skin after wounding. *Nature*, *447*(7142), 316–320.
Ito, T., et al. (2008). Immune privilege and the skin. *Current Directions in Autoimmunity*, *10*, 27–52.
Iwasaki, A., & Medzhitov, R. (2015). Control of adaptive immunity by the innate immune system. *Nature Immunology*, *16*(4), 343–353.
Jaeger, B. N., et al. (2012). Neutrophil depletion impairs natural killer cell maturation, function, and homeostasis. *The Journal of Experimental Medicine*, *209*(3), 565–580.
Jameson, J. M., et al. (2004). Regulation of skin cell homeostasis by gamma delta T cells. *Frontiers in Bioscience*, *9*, 2640–2651.
Jensen, K. B., & Watt, F. M. (2006). Single-cell expression profiling of human epidermal stem and transit-amplifying cells: Lrig1 is a regulator of stem cell quiescence. *Proceedings of the National Academy of Sciences of the United States of America*, *103*(32), 11958–11963.
Jensen, K. B., et al. (2009). Lrig1 expression defines a distinct multipotent stem cell population in mammalian epidermis. *Cell Stem Cell*, *4*(5), 427–439.
Jin, S., et al. (2021). Inference and analysis of cell-cell communication using CellChat. *Nature Communications*, *12*(1), 1088.
Jones, D. L., & Wagers, A. J. (2008). No place like home: Anatomy and function of the stem cell niche. *Nature Reviews. Molecular Cell Biology*, *9*(1), 11–21.
Joost, S., et al. (2016). Single-cell transcriptomics reveals that differentiation and spatial signatures shape epidermal and hair follicle heterogeneity. *Cell Systems*, *3*(3), 221–237 e9.
Knudsen, N. H., & Lee, C. H. (2016). Identity crisis: CD301b(+) mononuclear phagocytes blur the M1-M2 macrophage line. *Immunity*, *45*(3), 461–463.
Kobielak, K., et al. (2007). Loss of a quiescent niche but not follicle stem cells in the absence of bone morphogenetic protein signaling. *Proceedings of the National Academy of Sciences of the United States of America*, *104*(24), 10063–10068.
Komi, D. E. A., Khomtchouk, K., & Santa Maria, P. L. (2019). A review of the contribution of mast cells in wound healing: Involved molecular and cellular mechanisms. *Clinical Reviews in Allergy and Immunology*.
Kretzschmar, K., & Watt, F. M. (2014). Markers of epidermal stem cell subpopulations in adult mammalian skin. *Cold Spring Harbor Perspectives in Medicine*, *4*(10).
Lane, S. W., Williams, D. A., & Watt, F. M. (2014). Modulating the stem cell niche for tissue regeneration. *Nature Biotechnology*, *32*(8), 795–803.
Lee, J., et al. (2018). Hair follicle development in mouse pluripotent stem cell-derived skin organoids. *Cell Reports*, *22*(1), 242–254.
Li, L., & Xie, T. (2005). Stem cell niche: Structure and function. *Annual Review of Cell and Developmental Biology*, *21*, 605–631.
Li, C., et al. (2023a). An umbrella review of the use of platelet-rich plasma in the treatment of androgenetic alopecia. *Journal of Cosmetic Dermatology*, *22*(5), 1463–1476.

Li, J., et al. (2023b). Dermal PapillaCell-derived exosomes regulate hair follicle stem cell proliferation via LEF1. *International Journal of Molecular Sciences, 24*(4).
Liang, Y., et al. (2023). Adipose mesenchymal stromal cell-derived exosomes carrying MiR-122-5p antagonize the inhibitory effect of dihydrotestosterone on hair follicles by targeting the TGF-beta1/SMAD3 signaling pathway. *International Journal of Molecular Sciences, 24*(6).
Lien, W. H., et al. (2014). In vivo transcriptional governance of hair follicle stem cells by canonical Wnt regulators. *Nature Cell Biology, 16*(2), 179–190.
Lim, X., et al. (2013). Interfollicular epidermal stem cells self-renew via autocrine Wnt signaling. *Science, 342*(6163), 1226–1230.
Lim, X., et al. (2016). Axin2 marks quiescent hair follicle bulge stem cells that are maintained by autocrine Wnt/beta-catenin signaling. *Proceedings of the National Academy of Sciences of the United States of America, 113*(11), E1498–E1505.
Linehan, J. L., et al. (2018). Non-classical immunity controls microbiota impact on skin immunity and tissue repair. *Cell, 172*(4), 784–796 e18.
Liu, X., et al. (2021). Distinct human Langerhans cell subsets orchestrate reciprocal functions and require different developmental regulation. *Immunity, 54*(10), 2305–2320.
Liu, M., et al. (2022a). Dendritic epidermal T cells secreting exosomes promote the proliferation of epidermal stem cells to enhance wound re-epithelialization. *Stem Cell Research & Therapy, 13*(1), 121.
Liu, M., et al. (2022b). Intrinsic ROS drive hair follicle cycle progression by modulating DNA damage and repair and subsequently hair follicle apoptosis and macrophage polarization. *Oxidative Medicine and Cellular Longevity, 2022*. 8279269.
Liu, Z., et al. (2022c). Glucocorticoid signaling and regulatory T cells cooperate to maintain the hair-follicle stem-cell niche. *Nature Immunology, 23*(7), 1086–1097.
Liu, S., et al. (2023). Biosensors integrated 3D organoid/organ-on-a-chip system: A real-time biomechanical, biophysical, and biochemical monitoring and characterization. *Biosensors & Bioelectronics, 231*, 115285.
Lu, C. P., et al. (2012). Identification of stem cell populations in sweat glands and ducts reveals roles in homeostasis and wound repair. *Cell, 150*(1), 136–150.
MacLeod, A. S., & Mansbridge, J. N. (2016). The innate immune system in acute and chronic wounds. *Advances in Wound Care (New Rochelle), 5*(2), 65–78.
Maricich, S. M., et al. (2009). Merkel cells are essential for light-touch responses. *Science, 324*(5934), 1580–1582.
Martin, C., et al. (2003). Chemokines acting via CXCR2 and CXCR4 control the release of neutrophils from the bone marrow and their return following senescence. *Immunity, 19*(4), 583–593.
Martinez, F. O., & Gordon, S. (2014). The M1 and M2 paradigm of macrophage activation: Time for reassessment. *F1000Prime Reports, 6*, 13.
Matsumura, H., et al. (2016). Hair follicle aging is driven by transepidermal elimination of stem cells via COL17A1 proteolysis. *Science, 351*(6273), aad4395.
Mendez-Ferrer, S., et al. (2008). Haematopoietic stem cell release is regulated by circadian oscillations. *Nature, 452*(7186), 442–447.
Merad, M., Ginhoux, F., & Collin, M. (2008). Origin, homeostasis and function of Langerhans cells and other langerin-expressing dendritic cells. *Nature Reviews. Immunology, 8*(12), 935–947.
Meyer, K. C., et al. (2008). Evidence that the bulge region is a site of relative immune privilege in human hair follicles. *The British Journal of Dermatology, 159*(5), 1077–1085.
Mimeault, M., & Batra, S. K. (2010). Recent advances on skin-resident stem/progenitor cell functions in skin regeneration, aging and cancers and novel anti-aging and cancer therapies. *Journal of Cellular and Molecular Medicine, 14*(1–2), 116–134.

Mocsai, A. (2013). Diverse novel functions of neutrophils in immunity, inflammation, and beyond. *The Journal of Experimental Medicine, 210*(7), 1283–1299.

Moodycliffe, A. M., et al. (2000). Immune suppression and skin cancer development: Regulation by NKT cells. *Nature Immunology, 1*(6), 521–525.

Moussai, D., et al. (2011). The human cutaneous squamous cell carcinoma microenvironment is characterized by increased lymphatic density and enhanced expression of macrophage-derived VEGF-C. *The Journal of Investigative Dermatology, 131*(1), 229–236.

Muneeb, F., Hardman, J. A., & Paus, R. (2019). Hair growth control by innate immunocytes: Perifollicular macrophages revisited. *Experimental Dermatology, 28*(4), 425–431.

Nguyen, A., et al. (2018). Single cell RNA sequencing of rare immune cell populations. *Frontiers in Immunology, 9*, 1553.

Niederkorn, J. Y., & Stein-Streilein, J. (2010). History and physiology of immune privilege. *Ocular Immunology and Inflammation, 18*(1), 19–23.

Nouh, A. H., Abdelaal, A. M., & Fathy, A. E. S. (2023). Activated platelet rich plasma versus non-activated platelet rich plasma in the treatment of alopecia Areata. *Skin Appendage Disorders, 9*(1), 42–49.

Noy, R., & Pollard, J. W. (2014). Tumor-associated macrophages: From mechanisms to therapy. *Immunity, 41*(1), 49–61.

Parakkal, P. F. (1969). Role of macrophages in collagen resorption during hair growth cycle. *Journal of Ultrastructure Research, 29*(3), 210–217.

Parkinson, E. K. (1992). Epidermal keratinocyte stem cells: Their maintenance and regulation. *Seminars in Cell Biology, 3*(6), 435–444.

Paus, R., & Bertolini, M. (2013). The role of hair follicle immune privilege collapse in alopecia areata: Status and perspectives. *The Journal of Investigative Dermatology. Symposium Proceedings, 16*(1), S25–S27.

Paus, R., et al. (1994). Distribution and changing density of gamma-delta T cells in murine skin during the induced hair cycle. *The British Journal of Dermatology, 130*(3), 281–289.

Philpott, M. P. (2018). Culture of the human pilosebaceous unit, hair follicle and sebaceous gland. *Experimental Dermatology, 27*(5), 571–577.

Pillay, J., et al. (2012). A subset of neutrophils in human systemic inflammation inhibits T cell responses through Mac-1. *The Journal of Clinical Investigation, 122*(1), 327–336.

Pittet, M. J., et al. (2018). Recording the wild lives of immune cells. *Science Immunology, 3*(27), eaaq0491. https://doi.org/10.1126/sciimmunol.aaq0491.

Plikus, M. V., et al. (2008). Cyclic dermal BMP signalling regulates stem cell activation during hair regeneration. *Nature, 451*(7176), 340–344.

Portou, M. J., et al. (2015). The innate immune system, toll-like receptors and dermal wound healing: A review. *Vascular Pharmacology, 71*, 31–36.

Puga, I., et al. (2011). B cell-helper neutrophils stimulate the diversification and production of immunoglobulin in the marginal zone of the spleen. *Nature Immunology, 13*(2), 170–180.

Qian, C., et al. (2019). Heterogeneous macrophages: Supersensors of exogenous inducing factors. *Scandinavian Journal of Immunology*, e12768.

Qiu, J., et al. (2022). Efficacy assessment for low-level laser therapy in the treatment of androgenetic alopecia: A real-world study on 1383 patients. *Lasers in Medical Science, 37*(6), 2589–2594.

Rabbani, P., et al. (2011). Coordinated activation of Wnt in epithelial and melanocyte stem cells initiates pigmented hair regeneration. *Cell, 145*(6), 941–955.

Rajendran, R. L., et al. (2020). Macrophage-derived extracellular vesicle promotes hair growth. *Cell, 9*(4).

Rajendran, R. L., et al. (2021). Engineered extracellular vesicle mimetics from macrophage promotes hair growth in mice and promotes human hair follicle growth. *Experimental Cell Research, 409*(1), 112887.

Rak, G. D., et al. (2016). IL-33-dependent group 2 innate lymphoid cells promote cutaneous wound healing. *The Journal of Investigative Dermatology, 136*(2), 487–496.
Reichelt, J. (2007). Mechanotransduction of keratinocytes in culture and in the epidermis. *European Journal of Cell Biology, 86*(11–12), 807–816.
Ribatti, D., & Crivellato, E. (2012). Mast cells, angiogenesis, and tumour growth. *Biochimica et Biophysica Acta, 1822*(1), 2–8.
Rittie, L., et al. (2013). Eccrine sweat glands are major contributors to reepithelialization of human wounds. *The American Journal of Pathology, 182*(1), 163–171.
Rompolas, P., & Greco, V. (2014). Stem cell dynamics in the hair follicle niche. *Seminars in Cell & Developmental Biology, 25-26*, 34–42.
Rompolas, P., Mesa, K. R., & Greco, V. (2013). Spatial organization within a niche as a determinant of stem-cell fate. *Nature, 502*(7472), 513–518.
Rork, J. F., Rashighi, M., & Harris, J. E. (2016). Understanding autoimmunity of vitiligo and alopecia areata. *Current Opinion in Pediatrics, 28*(4), 463–469.
Schaerli, P., et al. (2004). A skin-selective homing mechanism for human immune surveillance T cells. *The Journal of Experimental Medicine, 199*(9), 1265–1275.
Scharschmidt, T. C., et al. (2017). Commensal microbes and hair follicle morphogenesis coordinately drive Treg migration into neonatal skin. *Cell Host & Microbe, 21*(4), 467–477 e5.
Schofield, R. (1978). The relationship between the spleen colony-forming cell and the haemopoietic stem cell. *Blood Cells, 4*(1–2), 7–25.
Sellheyer, K., & Krahl, D. (2010). Blimp-1: A marker of terminal differentiation but not of sebocytic progenitor cells. *Journal of Cutaneous Pathology, 37*(3), 362–370.
Sharp, L. L., et al. (2005). Dendritic epidermal T cells regulate skin homeostasis through local production of insulin-like growth factor 1. *Nature Immunology, 6*(1), 73–79.
Sheng, Z., et al. (2009). Regeneration of functional sweat gland-like structures by transplanted differentiated bone marrow mesenchymal stem cells. *Wound Repair and Regeneration, 17*(3), 427–435.
Stahl, P. L., et al. (2016). Visualization and analysis of gene expression in tissue sections by spatial transcriptomics. *Science, 353*(6294), 78–82.
Streilein, J. W., Alard, P., & Niizeki, H. (1999). A new concept of skin-associated lymphoid tissue (SALT): UVB light impaired cutaneous immunity reveals a prominent role for cutaneous nerves. *The Keio Journal of Medicine, 48*(1), 22–27.
Sun, Q., et al. (2023). Dedifferentiation maintains melanocyte stem cells in a dynamic niche. *Nature, 616*(7958), 774–782.
Sunderkotter, C., Kalden, H., & Luger, T. A. (1997). Aging and the skin immune system. *Archives of Dermatology, 133*(10), 1256–1262.
Suzuki, S., et al. (1998). Localization of rat FGF-5 protein in skin macrophage-like cells and FGF-5S protein in hair follicle: Possible involvement of two Fgf-5 gene products in hair growth cycle regulation. *The Journal of Investigative Dermatology, 111*(6), 963–972.
Taams, L. S., et al. (2005). Modulation of monocyte/macrophage function by human CD4 +CD25+ regulatory T cells. *Human Immunology, 66*(3), 222–230.
Taira, K., et al. (2002). Spatial relationship between Merkel cells and Langerhans cells in human hair follicles. *Journal of Dermatological Science, 30*(3), 195–204.
Tan, Y., et al. (2019). Non-ablative radio frequency for the treatment of androgenetic alopecia. *Acta Dermatovenerologica Alpina, Pannonica et Adriatica, 28*(4), 169–171.
Tanimura, S., et al. (2011). Hair follicle stem cells provide a functional niche for melanocyte stem cells. *Cell Stem Cell, 8*(2), 177–187.
Tigelaar, R. E., & Lewis, J. M. (1995). Immunobiology of mouse dendritic epidermal T cells: A decade later, some answers, but still more questions. *The Journal of Investigative Dermatology, 105*(1 Suppl), 43S–49S.

Tikhonova, A. N., et al. (2019). The bone marrow microenvironment at single-cell resolution. *Nature, 569*(7755), 222–228.
Tirosh, I., et al. (2016). Dissecting the multicellular ecosystem of metastatic melanoma by single-cell RNA-seq. *Science, 352*(6282), 189–196.
Toulon, A., et al. (2009). A role for human skin-resident T cells in wound healing. *The Journal of Experimental Medicine, 206*(4), 743–750.
Toyoda, M., Nakamura, M., & Morohashi, M. (2002). Neuropeptides and sebaceous glands. *European Journal of Dermatology, 12*(5), 422–427.
Trappmann, B., et al. (2012). Extracellular-matrix tethering regulates stem-cell fate. *Nature Materials, 11*(7), 642–649.
Uderhardt, S., et al. (2019). Resident macrophages cloak tissue microlesions to prevent neutrophil-driven inflammatory damage. *Cell, 177*(3), 541–555 e17.
van der Fits, L., et al. (2009). Imiquimod-induced psoriasis-like skin inflammation in mice is mediated via the IL-23/IL-17 axis. *Journal of Immunology, 182*(9), 5836–5845.
Van Keymeulen, A., et al. (2009). Epidermal progenitors give rise to Merkel cells during embryonic development and adult homeostasis. *The Journal of Cell Biology, 187*(1), 91–100.
Vollmers, A., et al. (2012). Two- and three-dimensional culture of keratinocyte stem and precursor cells derived from primary murine epidermal cultures. *Stem Cell Reviews, 8*(2), 402–413.
Wahle, P., et al. (2023). Multimodal spatiotemporal phenotyping of human retinal organoid development. *Nature Biotechnology*.
Wang, N., Liang, H., & Zen, K. (2014). Molecular mechanisms that influence the macrophage m1-m2 polarization balance. *Frontiers in Immunology, 5*, 614.
Wang, X., et al. (2017). Macrophages induce AKT/beta-catenin-dependent Lgr5(+) stem cell activation and hair follicle regeneration through TNF. *Nature Communications, 8*, 14091.
Wang, E. C. E., et al. (2019). A subset of TREM2(+) dermal macrophages secretes Oncostatin M to maintain hair follicle stem cell quiescence and inhibit hair growth. *Cell Stem Cell, 24*(4), 654–669 e6.
Wang, W. B., et al. (2023). Developmentally programmed early-age skin localization of iNKT cells supports local tissue development and homeostasis. *Nature Immunology, 24*(2), 225–238.
Watanabe, M., et al. (2017). Type XVII collagen coordinates proliferation in the interfollicular epidermis. *eLife*, 6.
Wojtowicz, A. M., et al. (2014). The importance of both fibroblasts and keratinocytes in a bilayered living cellular construct used in wound healing. *Wound Repair and Regeneration, 22*(2), 246–255.
Yang, S., et al. (2015). Immune tolerance. Regulatory T cells generated early in life play a distinct role in maintaining self-tolerance. *Science, 348*(6234), 589–594.
Yang, H., et al. (2017). Epithelial-mesenchymal micro-niches govern stem cell lineage choices. *Cell, 169*(3), 483–496 e13.
Yang, G., et al. (2023). Injury-induced interleukin-1 alpha promotes Lgr5 hair follicle stem cells de novo regeneration and proliferation via regulating regenerative microenvironment in mice. *Inflammation and Regeneration, 43*(1), 14.
Zhao, Q., et al. (2012). Activated CD69+ T cells foster immune privilege by regulating IDO expression in tumor-associated macrophages. *Journal of Immunology, 188*(3), 1117–1124.
Zhao, Y., et al. (2018). The origins and homeostasis of monocytes and tissue-resident macrophages in physiological situation. *Journal of Cellular Physiology, 233*(10), 6425–6439.

Printed and bound by CPI Group (UK) Ltd, Croydon, CR0 4YY
08/05/2025
01864965-0002